Repowering Communities

Repowering Communities

Small-Scale Solutions for Large-Scale Energy Problems

Prashant Vaze and Stephen Tindale

earthscan
from Routledge

First published 2011
by Earthscan
2 Park Square, Milton Park, Abingdon, Oxon OX14 4RN

Simultaneously published in the USA and Canada
by Earthscan
711 Third Avenue, New York, NY 10017

Earthscan is an imprint of the Taylor & Francis Group, an informa business

British Library Cataloguing in Publication Data
A catalogue record for this book is available from the British Library

Library of Congress Cataloging in Publication Data
Vaze, Prashant.
   Repowering communities : small-scale solutions for large-scale energy problems /
Prashant Vaze and Stephen Tindale.
      p. cm.
   Includes bibliographical references and index.
   ISBN 978-1-84971-266-8 (hardback) -- ISBN 978-1-84971-267-5 (pbk.) 1. Energy
policy. 2. Energy development. 3. Power resources. I. Tindale, Stephen. II. Title.
   HD9502.A2V39 2011
   333.79--dc22
                              2010053837

ISBN: 9781849712668 (hbk)
ISBN: 9781849712675 (pbk)

Typeset in Avenir and Caecilia
by Safehouse Creative

Contents

List of Figures, Boxes and Tables

Figures

Boxes

Tables

Foreword

Repowering Communities is an excellent and timely book that provides well-thought-out answers to how we can, and must, transform the way we generate and use energy if we are to overcome climate change and create sustainable employment in the 21st century. It is a must-read for everyone who wants innovative ideas about solutions to our energy challenges. It is also written in an entertaining and informative style – it is a book you just cannot put down – something more usually said about novels; I read it in one sitting.

The critical part of this book is the absolute clarity with which Prashant Vaze and Stephen Tindale present the need for a radically different model of energy production and distribution – one grounded in communities and the idea that, wherever possible, energy should be produced near to where it is consumed, allowing for greater efficiency, less waste and more inclusion of low-carbon solutions in the energy-generating mix. Their arguments are very consistent with the energy strategy the City of Toronto adopted while I was Mayor – a strategy based in conservation, demand management and distributed energy through a smart grid. In Toronto's case, the prediction of our wholly government-owned distribution utility, Toronto Hydro, was that this strategy would allow the city, its residents and businesses to meet or exceed Kyoto targets for greenhouse gas reductions.

The book is full of clear insights that lead inevitably to this conclusion. For example, early on the authors point out that it is very unlikely that a private utility, which exists to maximize value to its shareholders, will invest significantly in conservation. The business mission of such a utility is to sell more energy, not less – they are precisely the wrong institution to deliver such programmes – it is a bit like asking GM or Ford to fund public transit to ease traffic congestion. If we are going to encourage innovation in conservation and demand management, we need different actors involved in the market – and a persuasive and powerful case is made by the authors that locally based ESCOs are the right vehicle. They are the first authors I have read to understand the potential of engaging cities and communities, not just large utilities, as providers of energy solutions.

Do I agree with every single part of the book? Of course not. The authors are optimistic about the unproven carbon capture and storage. But the beauty of the book is that it is not necessary to agree with every detail, because they have the big picture right in a profoundly important way.

Prashant Vaze and Stephen Tindale make a strong call for political leadership. By charting a clear and credible path forward, they have made it easier for elected officials to take that leadership, and I am certain we will see it – not the least from the mayors of leading cities of the world. For that, the authors deserve our thanks – and congratulations.

David Miller
Mayor, 2003–2010, Chair of C40 cities, 2008–2010
Toronto

Acknowledgements

Stephen and Prashant first discussed the ideas in this book in 2006 when Stephen led Greenpeace UK and Prashant worked as an economist in the UK Department of Environment, Food and Rural Affairs. The idea of decentralized energy was beguilingly attractive, but never received political lift-off. Somehow the energy people in government were always a bit sniffy about small-scale energy and never saw it as properly serious compared to building nuclear reactors or pouring billions more into transmission or distribution lines. Nonetheless, the 2006 Energy White Paper had a chapter on it and energy efficiency; the Heat call for evidence published in 2008 featured a lot of the ideas too but progress in policy, except for the feed-in tariff, has been muted and delivery on the ground even less impressive.

Early in 2010 we decided to write a book that would showcase some of the successful projects around the world. It's easy to overlook the huge number of successful community and city initiatives that are taking place in North America and mainland Europe. By their nature many of these schemes are central to the operation of a city or a region but are not able to project this success to an international audience.

This was the task we set ourselves. The book has also been terrific fun to write. It's given us a licence to go and speak with interesting people about an issue we find fascinating. Thanks go to Peter Meyer who started the journey with us, but his work commitments meant he couldn't join us to the ultimate destination. Some of the research in this book is taken from articles Peter wrote for the Climate Answers website. Kristen Yount also made a huge contribution. Thanks also to Ed Mayo – the germs of many of the ideas in this book were worked on in a paper Ed and Prashant published in November 2009.

Many people have given generously of their time, commented on drafts and fed in ideas or data. In this list we include Maya de Souza, Prabhat Vaze, Ed Mayo, Malcolm Rigg, Liz Lainé, Richard Hall, the Serious Book Club, Graham Meeks, Michael King, Michael Felix, David Green, Larry Whitty and Ian Preston. Prashant spent a fascinating month in the US and Canada. People he'd like to acknowledge are, from Ontario: Paul Parker, Mary-Jane Patterson, Andrew Wilcox, Noel Cheeseman, Richard Morris; from Massachusetts: Sue Coakley, Audrey Schuman, Donald Kelley, Morris Pierce; from Vermont: Blair Hamilton and staff from Efficiency Vermont, David Farnsworth and others at Regulatory Assistance Project, Tom Stevans, Randy Pratt, Paul Zabriskie, R. Scott Campbell, Liz Schlegel;

from California: Mark Toney, Billi Romain, Jeff Shields, Jim Barnett, Wade Hughes, Ed Humzowee, Dawn Weisz, Devra Wang, Meg Gottstein, Pat Stoner, Rich Brown and Paul Frankel. We also had excellent and useful discussions with Simon Reddy, director of the C40 group, plus Alexandra Van Huffelen, Katinka Barysch, Simon Roberts, Birgir Lauersen, Marian Spain, Abigail Burridge, Ingrid Holmes, Simon Tilford, Andrew Warren, and Joel Kendrick. Staff at Earthscan have, again, been wonderful, especially Liz Riley, Hamish Ironside, Alice Aldous and Michael Fell.

Prashant Vaze and Stephen Tindale
May 2011

List of Acronyms and Abbreviations

°C	degrees Celsius
AC	alternating current
AD	anaerobic digestion
CARE	California Alternate Rates for Energy
CCS	carbon capture and storage (or sequestration)
CDU	Christian Democratic Union
CEGB	Central Electricity Generating Board
CERT	Carbon Emissions Reduction Target
CFL	compact fluorescent light bulb
CHP	combined heat and power
CO_2	carbon dioxide
CPUC	California Public Utilities Commission
CSP	concentrated solar power
CVCAC	Central Vermont Community Action Committee
DC	direct current
DH	district heating
DSM	demand-side management
EDF	Électricité de France
EECBG	Energy Efficiency and Conservation Block Grant
EPS	emissions performance standard
ESCO	energy services company
ETR	environmental tax reform
ETS	Emissions Trading Scheme
EU	European Union
EuroACE	European Alliance of Companies for Energy Efficiency in Buildings
FDP	Free Democratic Party
FiT	feed-in tariff
GDF	Gaz de France
GDP	gross domestic product
GW	gigawatt (1000MW)
HEET	Home Energy Efficiency Team
HVAC	heating, ventilation and air conditioning
IEA	International Energy Agency
IPCC	Intergovernmental Panel on Climate Change
kW	kilowatt
kWh	kilowatt hour
LADWP	Los Angeles Department of Water and Power
LED	light-emitting diode
LPG	liquefied petroleum gas

MASSIG	Market Access for Smaller Size Intelligent Electricity Generation
MW	megawatt (1000kW)
NIMBY	not in my backyard
NIMTO	not in my term of office
OECD	Organisation for Economic Co-operation and Development
PACE	Property Assessed Clean Energy
PFI	Private Finance Initiative
PG&E	Pacific Gas & Electric
PURC	public utilities regulatory commission
RAB	regulated asset base
RCI	Rotterdam Climate Initiative
REA	Rural Electrification Administration
RECO	Residential Energy Conservation Ordinance
REEP	Residential Energy Efficiency Project
RGGI	Northeast Regional Greenhouse Gas Initiative
RHI	Renewable Heat Incentive
RPS	Renewables Portfolio Standard
SCE	Southern California Edison
SDG&E	San Diego Gas & Electric
SIR	savings to investment ratio
SPD	Social Democratic Party
SMUD	Sacramento Municipal Utility District
SSJID	South San Joaquin Irrigation District
TURN	The Utilities Reform Network
UCTE	Union for Coordination of Transmission of Electricity
UN	United Nations
VAT	value added tax
VEIC	Vermont Energy Investment Corporation
WCI	Western Climate Initiative
ZED	zero emissions development

Chapter 1

Introduction

Anyone embarking on a book on energy policy has to feel daunted. A search on Amazon's US site uncovers over 10,000 books on energy and policy. This vast literature encompasses books on renewables and energy efficiency, the history of energy, economic modelling and financing of the energy industry, calls-to-arms to change the way we view energy, books debunking energy myths, worried treatises on the threat from China or Russia, books seeing the future as nuclear or the future as solar or seeing no future at all. Nor are governments shy about creating institutions to develop energy policy. All the world's largest countries have ministries of energy, independent energy regulators, large research communities and independent panels of advisers. Sitting above, or perhaps to the side of, these many bodies is the International Energy Agency (IEA), which was established in the 1970s in the wake of the oil shocks to act as an energy policy adviser for its 28 member states. There is no shortage of advice or comment on energy policy.

So why did we write this book? It is apparent to us the energy policies are insufficient, and energy companies are incapable of making the transition to a low-carbon, secure and affordable energy system. We write as two policy wonks who have worked in this area of policy for over ten years. The two of us have worked as senior policy advisers in government, as the executive director of a large environmental NGO, in the consumer movement and as a renewables analyst within the energy sector. We live in the United Kingdom, and have some familiarity with, and envy of, energy policy in other western European countries.

We believe that change is possible, but such are the lead times with large infrastructure projects that the transformation of energy politics, markets and organizations will take the next quarter of a century. We need to get our policies and institutional structures right urgently. We believe that a major part of this transformation will involve local government and community groups taking a leading role in making energy supply and use much more efficient – so increasing comfort and health, creating many new jobs and cutting emissions – and in delivering renewable energy. Here is a summary of our dream.

Daring to Dream

It is June 2034. Prashant is at home alone in London. He's lived in the same house for 25 years, which is now super-energy efficient. Stephen has moved to Keswick in the Lake District – partly to work with local councils and the National Park Authority to promote clean energy and partly to try and keep fit by walking up mountains. They've arranged to speak on skype-render.

Stephen patches into the call: 'Sorry I'm late. Just been watching the football'. 'There was a football match?', Prashant says blankly. He notices the dial on the energy display go red as the electricity tariff jumps to its maximum level. All around the country kettles, toasters, microwaves, the phone exchanges are swinging into action, craving power as people exit the football holograms. The grid balancing software will be assessing the spike in demand and sending out an offsetting flow of instructions: electric cars will flick from recharging their batteries to supplying power, washing machines will wait until the price of power drops, fridges will power down for a few minutes.

'Yes, England have won the World Cup', says Stephen, unable to control his delight (and not trying very hard to do so). 'Amazing. Proves that anything really is possible'.

Prashant tries to sound interested, but only perks up when the conversation switches to energy. 'So here we are, 25 years after deciding to write that book. Not everything we argued for has happened, but most of it has. Local councils and other community groups have led the energy revolution. National governments have woken up to the scale of the challenge – and the scale of the revenue they could get from taxing energy properly. The climate sceptics have realized that economies built on renewables are more successful than those based on fossil fuels from unfriendly parts of the world. Who'd have thought it?'.

Energy prices – even before the carbon and energy tax – have quadrupled as the demand from the Indian and Brazilian petrochemical industry gobbled up global gas output. Prashant's praise for local councils is in part based on his experience in Camden, London, where he lives. The council had set up an energy services company (ESCO) in 2012 initially as a contract with the local transition town to provide energy efficiency, but this brave initiative had grown and it now provided Camden residents with all their domestic energy needs. Residents pay the same flat fee (and any extra charges when they override the system defaults) based on their house and lifestyle choices.

Prashant's home is heated from a district heating (DH) system. The heat network had been very successful, enabling the council to decommission the residential gas supply in 2025. The hot water is sourced from a 1.5 gigawatt (GW) coal power plant near Didcot fitted with carbon capture and storage (CCS). In summer, power is provided solely by UK's 35GW hydro, wind and tidal power and by Britain's new nuclear plants – the fifth of which finally came on line last year but not before dragging its French parent generator company into receivership.

The hot water is supplied through a vast pipe laid on the bottom of Regent's Canal. The mutual that owns the canal also uses it to convey liquefied carbon dioxide (CO_2) from the power station through the carbon grid to disposal sites in the North Sea. There is a delicious irony to this; until the 1960s the canal was used to transport coal into London, now it was transporting waste gases and surplus heat from combusting coal. People fishing in the canal probably have no idea that just below the surface of the still waters super-insulated pipes carry CO_2 at −60 degrees Celsius (°C), just a few inches away from the DH pipes carrying water at 110°C. The power station at Didcot only operates in the winter months; its power is chiefly used for the seasonal demands from heat pumps in London's leafy suburbs and hot water for the heat networks that permeate inner London. The installation of the heat network had gone smoothly. The one in Prashant's street was installed in a week. Thames Water had won the franchise to install and operate the DH system throughout north London. In many ways it had been a no brainer since the company was anyway upgrading the drinking water network and it made sense to install the hot water network at the same time. Its robot moles had replaced the pipework without a trench being dug.

Staff from Camden ESCO had installed the heat exchanger and shown Prashant and his wife how to operate it. It was even more straightforward than the gas condensing boiler that it replaced. While they were there, the ESCO's staff went around the house and checked it for airtightness and used a thermal imager to find and rectify any deficiencies in the insulation. The heat exchanger hadn't worked properly at first but the ESCO fixed it within a week; perhaps the fact Prashant's wife is a local councillor and on the ESCO's management board might have helped.

Because the house is in a conservation area and shaded by tall buildings it had been impossible to install any external insulation at the front of the house, or any solar technology. The ESCO had fitted remote control circuits to the fridge and washing machine that allowed the appliances' operation to be governed remotely using information from Elexon (the company that administers the contractual codes and rules for settling financial flows between generators and suppliers) on the price of power. Some of the ESCO's suggestions and incentives had surprised Prashant; he was compensated £25 to move the fridge from the kitchen to a cooler location beneath the stairs. All the lights and monitors were replaced with light-emitting diodes (LEDs) and this had further reduced the house's already very low electricity use. Prashant gets the first 2000 kilowatt hours (kWh) a year of electricity included in his standing charge. He and his wife have never exceeded this limit. When he had the downstairs floor remodelled last year, the builder (certified by the ESCO) said there was a £1000 subsidy to build a lobby just inside the front door.

No one in the street knew or cared about how the power industry worked. The ESCO had taken over responsibility for energy supply when Camden implemented its powers to localize energy supply in 2017. The local authority had awarded the franchise to Camden Heat and Power, itself a commercial venture run by the transition town. The ESCO produces and dispatches electricity from its energy-from-waste plant in Wembley, which is co-owned with several other boroughs. Most of its electricity comes from the 'Big 8' energy generators. At night the generators idle their fossil and biomass plants. Intriguingly this means that only base-load power is dispatched from the centralized generators. At these times the cost of power is determined by the ESCO's demand-side management (DSM) decisions instead of the power station's 'merit order'. The man or woman in the street doesn't need to know this, but this flip in the market means that energy demand, not energy supply, has the upper hand in the power market. This reversal means that energy price is determined by a million demand-side decisions and not by the generators. This flip would have caused the whole energy market to judder to a halt but some wise decisions in the 2010s enabled the demand-side market to flourish. The fossil fuel generators' financial position had become more precarious as they were exposed to the uncertainties in fossil fuel prices. The financial markets created long-term contracts to help them plan and finance for future changes in fossil fuel and carbon market prices. The ESCOs now have the majority of the seats on Elexon and they have created a stable administered price of electricity, which works in their interests to provide a predictable revenue stream to pay their members and investors.

Stephen lives in a rented flat in Keswick, a town in the northern Lakes. He'd moved there in 2020, and was happy to rent because in 2015 the UK government started regulating the energy efficiency of the private rented sector. This was six years after the Swedish government had done the same, but to be only six years behind the Swedes wasn't too bad for British politicians of that era. Catching up with Scandinavia on carbon taxes had taken the UK a quarter of a century. Keswick council had installed solar thermal panels to heat hot water on most of the properties in the town, having worked out that solar photovoltaics for electricity generation are not ideal for northwest England but that solar thermal is highly efficient and economic even in the Lake District where, as the name suggests, there's a fair bit of rain. The Lake District National Park Authority has increased the amount of hydroelectricity produced, not by building reservoirs but through 'run-of-river' schemes that don't ruin the beautiful landscape. It has increased the use of waste wood from the forests for heating and electricity generation in decentralized combined heat and power (CHP) plants, with the heat distributed via small DH systems. A much larger DH scheme brings heat long distance from power plants on the Cumbrian coast to towns like Lancaster and Kendal.

The National Park Authority has consistently supported most (not all – some landscapes are simply too precious to alter) proposed wind farms in Cumbria, recognizing that the greatest threat to the landscape was uncontrolled climate change. The fact that a wind farm would be visible from inside the national park was not in itself reason to oppose it. On clear days (which do occur, even in the Lake District) many forms of human activity are visible from the mountaintops: cities, motorways and power stations. The local authorities around the Lake District also became supportive of wind farms, not least because from 2010 they had been able to make money by selling the electricity and from 2011 had got the business rates from any new wind farms in their area. Eden District Council, to the east of the national park, had lived up to its name by becoming one of the first areas to get all its energy from renewables.

The above account is of course science fiction. Neither of us expects England to win the World Cup within our lifetimes. The rest however is possible. Over the next few decades the UK, US and northwest Europe need to make some fundamental changes in their energy systems if they are to achieve their objectives of affordable warmth and security of supply. Many of the technologies we need have already been invented, the behaviours we need to display are known, the money to finance these changes is available in pension funds and savings accounts, and the political and social urge to make progress now is all too real.

We need a different approach to energy policy that sees the virtue of meeting energy service needs, of supplying sustainable energy, of moderating demand. We have to reach inwards to people and businesses to persuade them to change how they think about and use energy. How can existing homes and offices be adapted to dramatically reduce their energy use? How do we motivate people to make the day-to-day changes to their lifestyles to reduce their need for energy? This includes prosaic matters such as dressing appropriately for the climate, learning how to better operate their appliances, electronics and heating systems. It also raises uncomfortable questions such as should empty nesters move to smaller homes and what is the trade-off between conserving our built heritage and improving its energy efficiency? How do we obtain permission from our communities to site thermal power stations close to the communities, if we are to cheaply utilize the waste heat? How can we blend the facilities that provide energy into our landscapes to make use of the best renewable resources? These are changes that we cannot and should not expect distant companies and governments to make for us; they are changes that communities have to debate, reaching some form of qualified consensus, and then deliver for themselves.

No country or community has yet resolved such conundrums. But many communities, local and national governments are independently groping through the fog and are seeking solutions to these problems. We hope this book will inspire readers to believe solutions are possible. Examples of good practice have to become standard practice.

Competition or Co-operation

Someone living in Europe could be forgiven for thinking there was some unalterable law that means that the structure of the gas and electricity industry moves from public ownership to private ownership to liberalization. This view is very much encouraged by the European Union (EU), which has developed a series of directives to increase Europe-wide competition and liberalization as part of the EU Commission's wider agenda for pan-European economic integration. But in North America public policy has been more ambivalent towards the relative merits between private and public ownership. The energy ecosystem has always had greater diversity. Most states in the US experience a mixed ecology of private, public and co-operative ownership. Despite its reputation as the home of capitalism and free markets, the US is a leader in municipal and co-op energy firms. In Canada, provinces have established their own electricity generation businesses that are responsible for developing hydro (Ontario and Quebec) and thermal (nuclear, coal and gas) power plants. These businesses remain publicly owned monopolies, though there was a brief flirtation with retail competition in Ontario.

Before the 19th century, energy use in Western Europe and North America was decentralized and largely renewable. People used locally sourced wood for heating, wind and hydro supplied mechanical energy to mill grain and lift water, and whale oil and later kerosene were used to light homes and streets. Such decentralized energy didn't necessarily mean local sourcing – whales were often captured thousands of miles from where they were used; kerosene used in lamps across Europe might be manufactured in Poland. But individuals and communities could control and access energy without waiting for remote firms' decisions, though access was restricted to the wealthy.

This individual control meant there was more of a stewardship to energy management. Prashant remembers his childhood in India where a hot bath required water to be heated on a stove powered by liquefied petroleum gas (LPG). The process was intensely physical: the gas was conveyed to the home by bicycle rickshaw in huge metal pressurized canisters, the hot water was messily transferred to a bucket and then carried to the bathroom – a grossly inefficient mechanism for distribution and one that no one would like to see a return to. But the necessity of moderating demand was all too clear; the sweat on the rickshaw driver's face was a palpable reminder of the precious physicality of energy use. When he came to England and accessed hot water by simply turning a tap, this day-to-day reminder of scarcity was gone.

The Need for Change

Chapter 2 summarizes why transformation is needed. The main reason is the urgent need to radically reduce greenhouse gas emissions. Most scientists now agree that the world is warming and that human activity is very probably part of the cause. There are of course some scientists who dispute this, just as there are some who still dispute the theory of evolution. Science works through end-less questioning, so it is no bad thing that there are people who challenge the majority view. But this cannot be used as an excuse for inaction. The countries of northwest Europe and North America are not on track to reduce their greenhouse gas emissions by anything like the required amount.

Controlling climate change is not the only reason why transformation is need-ed. We are too reliant on energy imports from unstable and unreliable parts of the world. This issue is usually called 'energy security' by policy wonks. Both the US and northwest Europe are major importers of natural gas. New techniques to extract shale gas are allowing us to increase the amount of gas and beguiling the gas markets that are lowering world prices, but this technique does not make gas reserves any less finite.

Another issue making transformation essential is the fact that millions of peo-ple cannot afford to keep warm. Fuel poverty kills many thousands each year. As the world heats up, more will be threatened by heat waves – the 2003 heat wave in Europe killed 70,000 people – so air conditioning will become more a necessity than a luxury. Municipal and community energy organizations can deliver energy at lower cost than private companies can, in part because they have to pay less to shareholders.

Then there is economics. Billions of dollars, euros and pounds are wasted every year because so much energy is wasted. The heat from most power sta-tions simply goes up chimneys, so more fuel has to be burnt to warm homes and provide heat for industry. Then much of this heat leaks through the walls, lofts and windows of existing buildings. So even if you are a 'climate sceptic', there are very strong reasons why doing things differently is extremely sensible.

Progress So Far

We are convinced that energy transformation is not only necessary but also pos-sible. A number of communities have already made good progress. For this book we researched many different projects and interviewed some of the forces behind these initiatives.

Denmark is without doubt the country with the best-articulated policy to lo-calize its energy production. National initiatives (such as high energy taxes) and local regulatory powers mean that co-operatively and municipally owned energy suppliers thrive. We look at how Copenhagen developed its DH system and is now beginning to decarbonize the heat supplied to the system. We also look at the role played by customer-owned co-operatives that build and operate wind farms and small heat distribution systems.

Upper Austria has led the way in the use of solar thermal energy for space and water heating, heat pumps and biomass for DH. Berlin has shown that radi-cally improving the energy efficiency of existing buildings is both possible and affordable. Like Upper Austria it also utilizes biomass district heating.

In the US we look at Vermont's energy efficiency utility and a small farmer-owned electricity co-operative , which has rolled out smart metering through its customer base. In California, the City of Berkeley has introduced local ordinances mandating the installation and inspection of energy efficiency measures retrofitted into homes before they can be sold or extensively refurbished. In Sacramento, the municipal utility has for several decades been encouraging customers to deploy energy efficiency, shift their energy use to dampen peak demand and install photovoltaic panels. In Kentucky we look at ambitious and comprehensive programmes to improve school energy efficiency. Many states in the US have passed Property Assessed Clean Energy (PACE) legislation, which allows homeowners to borrow money from the local government, with the loan attached to their property rather than the individual. Interest is recouped through property tax.

Toronto in Canada is one of the leading cities globally, implementing numerous low-carbon policies to retrofit external insulation to large tower blocks and provide low-cost loans to improve energy use in municipal buildings, hospitals and universities. It operates a district cooling system by drawing water from the year-round cold depths of Lake Ontario.

In the UK, local governments in Woking and Southampton have established ESCOs and are developing DH systems powered by a mixture of high-efficiency CHP and geothermal energy. Aberdeen Council has also substantially expanded its DH programme.

More information about these local government and community initiatives is available on the 'Climate Answers' website (www.climateanswers.info).

Barriers to Progress

Sadly, these examples are far from the norm; progress is not nearly great enough nor fast enough. In Chapters 3 and 4 we consider why this is so. Chapter 3 looks at why the companies that dominate today's energy system are not delivering the transformation required. Privately owned energy companies make more money the more energy they sell and so are not very serious about energy efficiency. The operations of some of the large European utilities span many different territories. These firms, though once publicly owned, are now largely untamed. Strategic decisions on where and when to invest are made on commercial grounds, weighing up different country's incentives and their own desire to maintain a diverse portfolio of projects. In the US, the state regulation of utilities has produced a greater diversity of companies.

In Chapter 4 we consider why politicians and the policies they implement are not delivering, either at national or local levels. There is no shortage of energy policies, targets and timetables. Some of the goals – such as energy security – have the broad support of the whole population, while others – notably the need to reduce greenhouse gases – are disputed within communities. There are tensions between some of these policies, and differences in emphasis between countries, within countries and also between successive administrations. As a result, energy policy is always changing, undermining much of the long-term investment that is necessary.

How to Achieve Energy Transformation

Chapter 5 looks at the alternative structures of successful energy companies and ESCOs. These are engaged day to day in delivering sustainable energy. To do this they have secured the trust of their communities and started to make the necessary changes. For instance, changing the external look of homes or the siting of local energy facilities are among the most contested alterations one can make to a community. Organizations that have negotiated such changes include dedicated energy service or energy efficiency companies, businesses owned by local government, and co-operatives owned by the customers themselves. These business models must become far more widespread.

Chapter 6 considers where the money needed for investment in energy efficiency and community energy should come from. Developing decentralized energy sources and retrofitting homes and offices means mobilizing many tens of billions of pounds, dollars or euros of investment. The rates of return from energy efficiency and local renewables are low and seen as risky by the investment community. We consider the different financing mechanisms local governments, businesses and homeowners are using to access finances, and the compromises that property owners or their agents will have to make to reduce the risk to investors providing loans.

Chapter 7 covers energy taxes, incentives for renewables and tariff structures. Taxes on domestic energy have been a political no-go area in the US and the UK, but are widely used in Scandinavia and Germany. We assess energy taxes and subsidies for low-carbon changes to homes. Legitimate concerns about the affordability of warmth for the poorest sections of our societies have blinded us to the reality that for the richest half of our society the cost of energy barely registers as an issue. Domestic energy taxes are an essential mechanism for moderating demand and providing resources to governments.

Chapter 8 considers how citizens and communities can be motivated to change their behaviour to reduce energy use. The development of new energy businesses and better public policy will not remove the need for all individuals to change how they use their homes. Why do we change our behaviour? Who do we trust to inform us? How do we learn?

Chapter 9 draws these different threads together. We make ten recommendations that we think are needed to transform energy markets and allow much more bottom-up participation.

Communities and local government need to become much more active in delivering energy services, and especially energy efficiency. They cannot do all that is necessary. Large utilities are also needed, to continue their existing role in building large, centralized low-carbon energy facilities. To take one example: the energy company E.ON is investing in a 1GW offshore wind farm in the Thames Estuary, the London Array. This is very much part of our future. It would not be practical to expect London local government or communities to embark on such projects. Repowering communities in the ways outlined in this book is part of the answer, but not the whole answer.

Nevertheless, getting communities enthusiastic about where their energy comes from and how they use it is essential. Without community and local government involvement, sufficient clean energy will never be constructed and energy will never be used efficiently. There are also non-energy benefits. Persuading people to be active in their community will provide more local employment opportunities, improve community spirit and lead to better services. These social and political aspects are part of the dream too.

Chapter 2

The Need for Transformation

*Faced with the choice between changing one's mind and proving that
there is no need to do so, almost everyone gets busy on the proof.*
John Kenneth Galbraith[1]

A transformation of the world's energy system is urgently required. But before
we talk about the challenges, we should say a word or two about the successes
of the existing energy system. The electricity and gas industry has developed so
extensive a network over the past century that almost everyone in North America
and Western Europe has access to electricity, and most people can access mains
gas. The system of regulating monopolies prevents utilities from nakedly over
charging consumers at the aggregate level, while many of the local air quality
problems that used to kill thousands of people are now behind us.

However, major change is needed fast for several reasons. The existing approach does not provide energy to all of the world's growing population. It does
not provide 'energy security', through which countries and communities are able
to meet their energy needs from within their own territories or from other friendly
parts of the world. Even in rich, developed countries such as the US or UK, millions of people cannot afford to keep warm. Yet energy is provided and then
used very inefficiently. The majority of the energy released when fuel is burnt to
generate electricity is simply wasted up cooling towers or dumped into the sea.

Above all, we need to stop burning fossil fuels in ways that damage the climate. Fossil fuels are not renewable and so will eventually run out, but this is not
the main reason why we must stop burning them. We hear lots about peak oil and
a bit about peak gas, but nothing about peak coal. Proven coal reserves are enormous. If all the proven fossil fuels are burnt without the carbon being captured
and stored, the global climate will spiral out of control. The main reason why energy transformation is essential is to radically reduce greenhouse gas emissions.

Climate Change Must Be Controlled

Most scientists now agree that the world is warming and that human activity is very probably part of the cause. There are of course some scientists who dispute this, just as there are some who still dispute the theory of evolution. Science works through endless questioning, so it is no bad thing that there are people who challenge the majority view. But this cannot be used as an excuse for inaction. It is not 100 per cent certain that human activity is warming the globe. The majority scientific view is that it is 90 per cent likely. Would you go on a journey if told that there was a 90 per cent chance of a crash before arrival? We definitely wouldn't.

Climate change is the clearest example of the limits of the term 'environmentalism'. Yes, if uncontrolled it will be disastrous for trees, birds and monkeys. But it will also be disastrous – indeed lethal – for millions of people. Already the increase in global temperatures, with consequent droughts, more extreme weather and the spread of tropical diseases such as malaria, is causing the death of 160,000 people each year. This is according to the Intergovernmental Panel on Climate Change's (IPCC) 2007 report.[2]

Any reference to the IPCC is guaranteed to get opponents of climate action going: this is after all a socialist political body bent on world government and the abolition of capitalism. Well, perhaps. It was created in 1988 by the United Nations (UN), with the support of those notable socialists Ronald Reagan and Margaret Thatcher. It is political in that its major reports have to be agreed by governments. That's why it's called the Inter*governmental* Panel. It isn't perfect and does make mistakes – the people drafting the reports are after all human. There is no such thing as consensus or universal agreement among scientists. Questioning and challenging are the very essence of science. It was a good marketing move of those opposed to reducing emissions to get themselves called 'climate sceptics'. Scepticism is the very essence of serious science. These people are not genuinely sceptical – they are opposed to the very notion of taking climate change seriously. The UK media uses the term 'Eurosceptics' in a similar way, to describe people who are opposed to European co-operation. What the IPCC has done, painstakingly over 23 years, is to identify what *most* scientists agree about. Their Fourth Assessment Report states that it is unequivocal that the climate is now warming, and that it is very likely that human activity plays a role in this. It defines 'very likely' as over 90 per cent probable. So it isn't certain but very probable. It isn't certain that your house will catch fire, and not even probable, but most people think it wise to take some precautions and buy some insurance.

In 2009 the Global Humanitarian Forum, a think tank run by former UN Secretary General Kofi Annan, published a report saying that climate change is already responsible for 300,000 deaths a year and is affecting 300 million people.[3] It argues that increasingly severe heat waves, floods storms and fires will be responsible for as many as 500,000 deaths a year by 2030, making it the greatest humanitarian challenge the world faces.

For those who don't trust politicians or former UN officials, perhaps doctors and academics are more persuasive: 'Climate change is the biggest global health threat of the 21st century' – these are the opening words of a major report from the Lancet and University College London, published in 2009.[4] Increasing global temperatures will lead to increased famine, spread of tropical diseases and death through extreme weather events. More than a sixth of the world population live in areas reliant on glacial meltwater, which will diminish or disappear totally. Poor countries and people are not the only ones threatened – we all are – but as usual the poor will suffer more.

The Lancet and University College London report goes on to say that much needs to be done by communities and local government to prepare for now inevitable climate change – adaptation is the policy jargon for this. But much can also be done to control climate change by reducing emissions – mitigation.

A lot of this mitigation can be achieved by using technology and different fuels. Coal has a very high carbon content. Generating electricity from coal causes over twice as much CO_2 per unit of electricity as from gas, eight times as much as nuclear, ten times as much as solar and forty times as much as wind.[5] The US gets about half of its electricity from coal, as does Germany. The UK gets about 40 per cent. And major developing nations are adding rapidly to the climate impact of coal. China gets 80 per cent of its electricity from coal and India 70 per cent. South Africa gets over 90 per cent. There is, understandably, much discussion of nuclear waste. There is much less discussion of coal waste, some of which remains on the ground but most of which is released into the atmosphere. It is possible to capture the CO_2 from coal and gas power stations and then store it in old oil or gas fields or saline aquifers. CCS (carbon capture and storage/sequestration) is proven at a small scale and at every stage of the process, but has yet to be demonstrated at a large scale or integrated throughout the process. There is no clear reason why it will not work, but it needs to be demonstrated. It will significantly increase the cost of generating electricity from coal and gas – though as the Stern Review (2006) memorably and correctly stated, controlling climate change will be cheaper than not controlling it.[6]

Energy is not just electricity. Much energy is used for transportation and almost all of this comes from oil at present, though transport is beyond the scope of the book. Much energy is also used for heating, which is central to this book. In the UK 47 per cent of annual greenhouse emissions come from fuel used for heat; in Germany this figure is 58 per cent and in the US 27 per cent. Reducing emissions from heat will come partly from using renewable gas rather than natural gas, or any gas rather than coal or oil. Greater reductions will come from insulating existing buildings and using the heat that is currently wasted when fuel is burnt to generate electricity.

According to the IPCC we need to reduce greenhouse gas emissions by around 80 per cent from 1990 levels by 2050.[7] Yet emissions from developed countries have risen since 1990, despite the Kyoto Protocol. Kyoto has achieved a little. In 2006 (so before the global recession) emissions from the countries that accepted targets and ratified them were 5.5 per cent lower than 1990 levels. EU emissions were 2.2 per cent lower than in 1990. But these figures include the purchase of emissions credits from countries such as Russia. Russia's Kyoto target was that emissions in 2010 should be the same as in 1990, but the collapse of much of the Russian economy meant that emissions in 2006 were in fact 34 per cent below 1990 levels. This enabled Russia to sell emissions credits to other countries, a process often (and legitimately) referred to as 'hot air'. The EU's Kyoto target would not be met without the use of hot air. Nevertheless, Europe did better than the US, which signed Kyoto but never ratified it. The US allowed its emissions to rise by 14.4 per cent by 2006.

Less Energy Must Be Imported from Unfriendly Parts of the World

The world is not short of fossil fuels. But unfortunately the plants and animals that died to become fossils and then fuels were not considerate enough to spread themselves evenly around the globe. So some countries have plenty of fossil fuel reserves while others have to import. This brings a high financial and diplomatic price. Even those people who dispute that burning fossil fuels causes climate change accept that having to be nice to regimes such as Saudi Arabia (relied upon by the US), Libya (by the UK) and Russia (by all of Europe) has some diplomatic downside and brings moral dilemmas. The US is a major importer of oil and gas (this mainly from Canada, where the diplomatic downsides are less serious). The UK was self-sufficient in oil and gas from the mid-1970s to the early 21st century but North Sea reserves are running out. Germany has plenty of coal but imports most of its gas from Russia. The US and Europe are becoming increasingly reliant on imported sources of energy. Fossil fuel production in the US and most of Western Europe has been falling for some years. New oil and gas fields are still being discovered but these discoveries are less than the decline in output of the larger spent fields.

Table 2.1 shows how the largest North American and European economies are now substantial net importers of energy. Norway, Canada and Denmark are net exporters, but the amount they export is dwarfed by the needs of their neighbours. As fossil fuel energy resources are depleted, more countries will move from being net exporters to net importers. After World War II the US was largely self-sufficient in crude oil. Domestic production was still rising until 1970. Since then it has dropped by a third.[8] In 1949 its net imports were just 6 per cent of consumption, and now net imports are 30 per cent. The UK had a brief flurry of exporting energy in the 1980s and 1990s as North Sea production came on stream. But production peaked in 1999 and the UK became a net importer of gas in 2004.[9]

Table 2.1 Total primary energy flows in 2007 (billion tonnes of oil equivalent)

	Consumption	Production	Net imports	Import dependency (%)
Canada	546	771	-225	-41
US	4030	2837	1192	30
Denmark	35	44	-9	-27
Germany	562	203	359	64
Netherlands	162	105	56	35
Norway	76	394	-318	-418
UK	375	292	83	22

Source: US EIA, www.eia.doe.gov/emeu/international/oiltrade.html

The UK now relies on imports by pipeline or gas tanker. But countries are all chasing supplies from the same group of energy exporters. Russia provides around a quarter of the natural gas consumed in Europe, most of it piped through countries such as Ukraine. In Europe we realized the risks this posed in the winter of 2005–2006 – Christmas is no fun if you are at the wrong end of a gas pipeline and the two countries the pipeline passes through are at loggerheads.

The only change in this narrative of fossil fuel decline has been the innovations in recovering gas from unconventional sources such as shale. Gas is recovered using techniques such as hydraulic fracking (liquids squirted into rocks at very high pressure to crack the rock) and sideways tunnelling several kilometres below the ground. This links together small pockets of gas deposits that wouldn't otherwise be economically exploitable. Present indications suggest this might increase the supply of gas substantially but the process is very expensive and there are environmental worries that the chemicals used for fracking harm the ground water, and about the full life-cycle greenhouse gas emissions of shale gas. Shale gas accounts for around 10 per cent of US production.

The effect of energy insecurity has been most obvious in the case of oil. The invasion of Iraq was not entirely an oil war: Saddam Hussein was a dangerous, genocidal dictator. But access to Iraq's oil was clearly a major factor. Of the five permanent members of the UN Security Council, Russia, China and France had signed post-sanctions agreements with the Saddam regime to enable them to buy the oil once sanctions were lifted. These countries opposed the invasion. The US and UK had not signed post-sanction agreements and made up the invasion force.

Energy Must Be Produced and Used More Efficiently

Billions of dollars are wasted every year because energy is wasted. The European Commission has estimated that reducing EU energy consumption by 20 per cent by 2020 would reduce the cost of energy imports by EU€100 billion to €150 billion annually and could create 1 million new jobs.[10]

Most of the energy from burning fuel in power stations simply goes up chimneys – this heat could potentially be used to warm homes and provide heat for industry. The majority of homes still have poorly insulated walls, lofts and windows, so the majority of energy used in buildings is for heating and cooling. So even if you are a 'climate sceptic', there are good reasons why doing things differently is extremely sensible. All countries need to focus on producing energy more efficiently and then using it more efficiently.

It is quite possible to use the heat from electricity generation for industrial or domestic heating. CHP (also known as cogeneration) is a well-established technology that makes obvious economic, energy security and climate sense. Yet less than 10 per cent of US electricity comes from CHP. (Even using the heat from this small proportion of power generation saves over $5 billion a year in energy costs, according to the US Clean Heat and Power Association.[11]) Only about 5 per cent of Canadian electricity comes from CHP plants.[12] Europe does better, but only slightly: in 2007, 11 per cent of EU electricity and 13 per cent of heat used came from CHP plants.[13]

Once energy is delivered to industrial and domestic consumers, it could and should be used much more efficiently. Efficient heat use in industry requires efficient machinery. In homes it requires energy-efficient buildings. Household use of heat and electricity is typically responsible for around a quarter of energy use in Europe,[14] commercial and public sector buildings use a further 13 per cent. Homes and businesses are responsible for around two-fifths of energy use.

There has been a steady improvement in the energy efficiency of new buildings. Well-built new buildings can be as comfortable as the old inefficient ones but use only a quarter of the energy for heating. The most efficient ones – the PassivHaus – use only about 10 per cent. But buildings last a long time. The reduction of energy wastage from buildings requires improvement of the existing stock, not just better new buildings.

Scandinavian homes 'have exemplary levels of residential energy efficiency'.[15] The Netherlands also has good buildings. Germany has relatively bad buildings (though the publicly owned KfW bank has been active in offering low-interest loans for energy efficiency improvements). But the most energy-inefficient buildings in northern Europe are in the UK and Ireland.

In Canada there have been efforts to create high-performance buildings, for example using the Energuide standard for homes, which was introduced in 1998, but the scheme was closed down in 2006 following budget cuts. In the US, the Department of Energy publishes and revises building energy codes for residential and commercial structures that states are expected to write into their own laws.

The fabric of the building is the key determinant of heat use but it is the behaviour of inhabitants and the products they use that are the key drivers of electricity use (at least in buildings that are not heated electrically). The trend towards bigger and more powerful electronic products means that the electricity consumption of European households and North American households is increasing, and the rate of increase in consumption has actually speeded up since 2000.[16]

Figure 2.1 shows that CO_2 emissions from energy used in the residential, commercial and institutional sectors fell between 1990 and 2008. Germany (–25 per cent), Sweden (–77 per cent) and Denmark (–35 per cent) have all recorded substantial reductions. Emissions have fallen slightly in the UK (–9.3 per cent) and US (–6.1 per cent) and risen in Canada. These results exclude emissions from generating electricity used in homes and commercial buildings. The data from the US have been divided and those from Denmark and Sweden multiplied by ten to fit on the same scale.

Not only are the Europeans making greater progress than the North Americans, they also started from much lower per capita greenhouse gas emissions across the whole economy. The amount of energy used per unit of gross domestic product (GDP), usually called energy intensity, varies substantially between economies, as the 2004 data in Table 2.2 show. Some of this is due to different latitudes – it is understandable that energy use in northern countries is

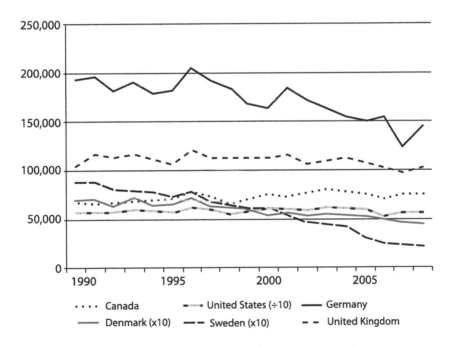

Figure 2.1 CO_2 emissions from fuel use in residential, commercial and institutional sectors since 1990 for selected countries (ktCO$_2$)

Note: Data from the US have been divided and those from Denmark and Sweden multiplied by ten to fit on the same scale.
Source: UNFCCC greenhouse gas inventory, www.unfccc.int/di/DetailedByParty/Event. do?event=go#

Table 2.2 *Energy use per unit of GDP and annual per capita CO_2 emissions, 2004*

Country	Energy use per unit of GDP (US = 100)	Annual per capita CO_2 emissions (tonnes/person)
Denmark	57	11.5
UK	62	11
Germany	75	11.9
Sweden	98	7.5
US	100	23.5
Finland	109	13
Canada	133	22.6

Source: UN for energy intensity data, www.un.org/esa/sustdev/natlinfo/indicators/isdms2001/isd-ms2001economicB.htm; World Resources Institute for per capita emissions, http://earthtrends.wri.org/searchable_db/index php?action=select_countries&theme=3&variable_ID=466

higher than in warmer southern ones. But Denmark shows that it is quite possible to be a northern country and still use energy efficiently and keep greenhouse gas emissions down. As is so often the case, and as we will frequently say in this book, other countries have much to learn from Denmark. (Sweden is doing better on per capita CO_2 emissions but this is due to its extensive use of hydroelectricity and the fact that it has kept its nuclear power stations running.)

The amount of energy used per unit of GDP declined on average 1.6 per cent a year between 1990 and 2006.[17] But these improvements in energy intensity have not led to a reduction in overall energy use. They have led instead to greater economic growth than would have been the case without better use of energy. Before the recession, energy use was increasing in most European countries (with the notable exception of Germany, where it was declining). Using energy more efficiently makes good economic sense, but if the result is that more energy is used, this does not help the climate. This 'rebound' effect is well established at the level of individual behaviour. A more efficient car becomes cheaper to run, leading to an increase in its use. The economist Jevons recognized this in the 19th century when he observed that technical innovation in steam engines improved the efficiency with which they used coal but paradoxically made coal more useful and so increased its use across society as a whole.

Energy Must Be Made More Affordable for the Poor, But More Expensive for the Rest of Society

Millions of people in Europe and North America cannot afford to keep their homes at a comfortable temperature. The inability to keep homes sufficiently warm, cool and free of damp kills many thousands each year. As the world heats, summer heat waves will threaten lives more – the 2003 heat wave in Europe killed 70,000 people.

In the UK over 4.5 million households in 2008 needed to spend more than 10 per cent of their income on keeping warm[18] and lit. Many of those vulnerable to fuel poverty are unemployed, disabled, ill or retired. These people spend more time at home and so use more energy, but have less income to pay for it.

Many environmentalists are pleased – either publicly or privately – when energy prices rise, since this means that people will be more careful how they use energy and become more motivated to install energy efficiency in their home. However, for people in fuel poverty, a sharp increase in energy prices or a cold winter can mean the difference between life and death. Most years, around 20,000 more people die in the winter months in the UK than in the summer months, in part because of respiratory diseases brought about from living in cold and damp homes. If there is an outbreak of influenza, the winter 'excess mortality' may easily be double this.

Fuel poverty is much less of a problem in Scandinavia, Germany, Austria or The Netherlands. These countries have a much more equal distribution of income and wealth than do the US and UK, and also have much better insulated buildings. This makes it politically possible and socially acceptable to increase the price of domestic energy through taxation. Due to taxation, there are very considerable differences in the energy prices paid by consumers in different European countries. Swedish consumers pay by far the highest gas prices, about 50 per cent more than Danish consumers, who face the second highest prices. British consumers pay the lowest gas prices of any EU member state. Denmark has the most expensive electricity tariff in the world. German household electricity tariffs are around two-thirds of Danish ones.

Domestic energy taxes have been off the political agenda in the UK and US for many years. In the UK a small tax on commercial and industrial energy use, the Climate Change Levy, was introduced by the Labour government. This has not been high enough to make individuals use energy more carefully in their offices. Consumer organizations such as Consumer Focus in the UK or the Utilities Reform Network in the US are fighting to keep energy prices affordable, arguing that price rises (which in the US have to be sanctioned by the energy regulators) are soft on the utilities and unfair to consumers, and especially to vulnerable consumers. Against them, scientists and climate campaigners say that the economy hasn't had a strong enough jolt through the rising energy prices to bring about real carbon saving, and that for most people prices are low enough to mean that only energy anoraks or budgeting obsessives bother to find out how much they pay for their energy.

In the US, the Housing and Urban Development Department has calculated what it calls the home energy affordability gap,[19] which is the amount by which energy bills exceed affordable energy costs. This was $34 billion in 2007–2008 at the average winter energy price. The government only allocated $1.9 billion that year. In 2002, the energy affordability gap was just $18 billion and the federal allocation was $1.7 billion. The gap is therefore getting much wider.

Electricity and gas are what economists call inferior goods. This means, that as families get richer, they spend a smaller proportion of their incomes on these goods. In 2008 it was estimated that some 4.5 million households in the UK (about one in six) needed to spend 10 per cent of their incomes on gas and power. In the US, Prashant saw areas in Vermont where some households spent 30 per cent of their incomes on keeping warm. Similar issues arise from the cost of keeping cool in inland areas of California. For these people the cost of maintaining their homes at a habitable temperature is an important issue, and at times of the year even a life and death issue.

But for the majority of people the cost of energy is not a burdensome part of their budget and simple energy-saving opportunities are overlooked. The gas and electricity bill is taken from their bank account every month, it's something they have no great control over and that's the end of it. The authors have interviewed friends and relatives – intelligent, well-educated professional people – who have no idea who supplies them with energy, no idea of its price and no idea of their monthly bill. For the richer half of UK society, gas and electricity only represented 4 per cent of their weekly spending in 2007[20] and these people account for 60 per cent of domestic energy use. This huge disparity of interests between different sections of our society often living cheek by jowl makes the issue of energy pricing and energy taxation a highly polarized debate.

It is not helped by the fact that few customers are presented with accurate bills in a timely manner. In most of Western Europe and North America customers are sent accurate bills every month. In the UK, meters only have to be read once a year so people receive correctly reconciled bills that accurately reflect usage a year late. People hope that smart meters will expose people to more accurate and timely information on their energy use. We discuss this in later chapters.

But better information, through better billing or public information campaigns, will never be sufficient to make organizations and individuals use energy more sensibly. Carbon or energy taxes are also necessary to make people and businesses pay more attention to their energy us. Taxes are also needed to promote cleaner energy sources and penalize dirty, inefficient fossil fuel plants. The development of new energy businesses that take over the energy management of homes and offices and make money from installing energy efficiency, which is central to the transformation this book is arguing for, will only happen if energy tariffs increase.

In a previous book[21] Prashant calculated the cost-effectiveness of carrying out various retrofits in his 19th century house. The payback periods of these changes were between 6 months and 200 years. This is in part because domestic gas is virtually untaxed. The UK has no domestic energy tax, the sales tax is applied at reduced rate and the fixed costs of the gas distribution system are recovered from a standing charges rather than the price of the last unit of gas being bought. Many green-minded consumers do similar sums for themselves and reach the same depressing realization: under current market prices there's no economic reason to do energy efficiency work. Changing a boiler, paying for professional draught proofing or fitting under-floor insulation only pays after many years. Why should the average person care?

The wish to keep the price of energy affordable is understandable, since so many low-income households are struggling to keep their homes warm. But a result is that the middle class and wealthy blithely use energy profligately. Politicians are well aware of the high cost of energy to some of their voters. Economists' jargon about internalizing externalities doesn't wash with the general population who just view it as more taxation. But we don't think that the correct response to helping the fuel poor is to keep all energy prices low for everyone. Higher energy taxes are essential to increase the cost of units of energy used and so make energy efficiency more economically appealing. The politics of energy taxes – in the Anglo-Saxon world at least – are pretty ferocious, and are considered in Chapter 4. But higher taxes are necessary to achieve energy transformation, and have the added bonus of raising enormous sums of money for governments, who could at present do with the extra cash. These should be introduced alongside rising block tariffs that keep the cost of the household's initial gas and electricity units reasonably low, but make the cost of units from each block above this considerably higher. This is explained in Chapter 7.

Tackling the global demand for energy without destroying climate stability needs two major changes. First, there needs to be a shift to much cleaner sources of energy. Wind, solar, wave, tidal and biomass (not based on cutting down trees) can eventually provide all the world's energy needs. But less than a tenth of total energy used currently comes from such sources, so it will take many decades to get to 100 per cent. This makes other low-carbon energy sources necessary as bridge technologies. Second, we need a shift from the 'predict-and-provide' business model in which governments and energy companies see their role as simply being to 'keep the lights on', delivering as much energy, wherever and whenever customers request. In future energy companies must become ESCOs, delivering light and warmth, not simply fuel. This will involve them doing much more on energy efficiency. Chapter 3 explains why the existing business model of energy companies is not well suited to them becoming ESCOs, as energy efficiency is not a 'product' that can be easily bought and sold. Community energy companies offer a much better business model: this is the subject of Chapter 5.

This book tries to offer suggestions on how the transformation can be achieved. But before we move to suggested solutions, we need to spend a bit more time on the problems. The next chapter outlines why existing energy companies aren't delivering energy efficiency and affordable clean energy. Chapter 4 covers why existing politicians and policies aren't delivering.

Notes

1   Quote widely attributed to economist Galbraith, www.saidwhat.co.uk.
2   IPCC (2007) *Fourth Assessment Report*, IPCC.
3   Global Humanitarian Forum (2009) *Human Impact Report*, Global Humanitarian Forum.
4   Lancet/University College London (2009) *Managing the Health Effects of Climate Change*, Lancet/UCL.
5   UK Energy Research Centre (2006) *Response to Treasury Consultation on Carbon Capture and Storage*, UK Energy Research Centre.
6   Stern, N. (2006) *Stern Review on The Economics of Climate Change (pre-publication edition). Executive Summary*, HM Treasury.
7   IPCC (2007) *Fourth Assessment Report*, IPCC.
8   Data from the US Energy Information Administration, www.eia.doe.gov/emeu/international/oiltrade.html.
9   Wicks, M. (2009) 'Energy security: A national challenge in a changing world', www.decc.gov.uk/assets/decc/What%20we%20do/Global%20climate%20change%20and%20energy/International%20energy/energy%20security/1_20090804164701_e_@@_EnergysecuritywicksreviewBISR3592EnergySecCWEB.pdf.
10 European Commission (2005) *Doing More With Less: Energy Efficiency Green Paper*, European Commission.
11 US CHPA (no date) 'Clean heat and power basics', US CHPA, www.uschpa.org/i4a/pages/index.cfm?pageid=3283.
12 OECD/IEA (2008) *Combined Heat and Power: Evaluating the Benefits of Greater Global Investment*, OECD/IEA.
13 Cogen Europe, www.cogeneurope.eu.
14 European Commission (2007) *Energy – Overview and Security of Supply*, Energy Commission. See Figure 2.2.7 showing final energy consumption by sector for 2007. Households use 285 million tonnes of oil out of 1158 million tonnes of oil.
15 Healy, J. D. (2003) 'Housing conditions, energy efficiency, affordability and satisfaction with housing: A Pan-European analysis', *Housing Studies*, vol 18, no 3.
16 World Energy Council (2008) *Energy Efficiency Policies Around the World: Review and Evaluation*, World Energy Council.
17 World Energy Council (2008) *Energy Efficiency Policies Around the World: Review and Evaluation*, World Energy Council.
18 DECC (2010) *Annual Report on Fuel Poverty Statistics 2010*, www.decc.gov.uk/assets/decc/Statistics/fuelpoverty/610-annual-fuel-poverty-statistics-2010.pdf.
19 Fisher Sheehan Colton (2009) 'On the brink: 2008 The home energy affordability gap', *Public Finance and General Economics*, Belmont, MA.
20 Office for National Statistics (2008) *Family Spending*, Office for National Statistics, Palgrave MacMillan, London. Derived from Table A.8.
21 Vaze, P. (2009) *The Economical Environmentalist: My Attempt to Live a Low-Carbon Life and What it Cost*, Earthscan.

Chapter 3

Why Existing Energy Companies Aren't Delivering – Can a Shark Go Vegan?

> *A key condition to the Spalding gas consent is that sufficient space remains available adjacent to the power station to allow for the retrofit of carbon capture plant at a later date. Waste heat from the Spalding plant could also be harnessed by local users such as the community hospital, and the option for this has been left open.*
> Press release for Spalding gas-fired power station, 11 November 2010[1]

The above quote, from a press release for a new UK power station hints at the delicate negotiations taking place between government and firms wishing to build new electricity (and heat) plant. Governments are desperately in need of investment, not just to replace aging assets but also to adapt to a future with tighter caps on carbon emissions and scarcer fossil fuels. They are asking firms to get ready for a time in the future where there is infrastructure to store CO_2 emissions and use waste heat, but the cost of these enhancements has to be borne now in anticipation of something that may or may not come to pass. At the same time, government recognizes the most sensible way to deal with the need for replacement power stations is through reducing the overall need for electricity and reducing the amount of heat lost through leaky homes. Improving energy efficiency reduces the need for new gas storage facilities; it would also increase energy security by reducing the amount of gas that needs to be imported. And using less fossil fuel would obviously help the climate too. But energy companies and their shareholders aren't that keen on energy efficiency because it damages their bottom line.

This chapter is about who owns and makes decisions about the gas and electricity supplied to our homes. This is a far from static picture: changing political fashions have seen the industry move between private and public ownership and between integration and disintegration of electricity generation, distribution and supply. And the mission of energy companies is quietly, and not necessarily consensually, being changed from energy provider to agent of social and environmental change.

From the start of the 19th century we began to rely on networked gas and electricity to provide energy to our homes. One only has to walk around an old European home to see fireplaces from the coal era or electrical wiring fitted on over plaster walls, as though a mid-20th century afterthought. The change from old to new sources of energy was a profound one that led to the modern age, and also new styles of industrial organization and economic interaction between home and communities and the energy businesses. How much more real our emissions of CO_2 would seem if we all had to handle fossil fuels every day, as we used to when coal and firewood were handled on a daily basis.

Thomas Edison, famous for being a great inventor, was a shrewd businessman too. He set up the first investor-owned electricity distribution company in 1878, which went on to become General Electric. But it was Nikola Tesla, the quirky genius (who by all accounts was afraid of dirt and obsessed with the number three) that invented the alternating current (AC) system for distributing power that we use today. This was a superior method to the (low voltage) direct current (DC) system that was being championed by Edison and allowed power generation to be centralized and then distributed through wires to individual customers. Tesla briefly worked for Edison, but they fell out when Tesla debugged Edison's problematic DC system but was refused the generous bounty that had been promised him because, as Edison meanly put it, Tesla had failed to comprehend American humour and didn't appreciate the promise of generous recompense had actually *been a joke*. Tesla's AC distribution patents were acquired and exploited by George Westinghouse in 1888. The technology was an immediate hit and by 1889 the Westinghouse Electric Corporation was formed and the technology was being franchised out across the US, with 30 more lighting systems within a year. Many of the early power companies exploited local hydro sources; the Buffalo scheme was powered by the nearby Niagara Falls. Even now utilities in the US, such as Southern California Edison and American Electric Power, can trace their roots to the flurry of investment during the start of the 20th century. At its outset electricity was locally produced, often from renewable sources of energy.

The earliest power companies in the US were local private concerns. By the turn of the 20th century the limitations of private sector benevolence were already becoming clear as swathes of the US were excluded from access to power, since there was little money to be made from linking sparsely populated rural areas. It wasn't until the Great Depression in the 1930s that rural co-ops formed with a dual mandate to provide work and power to the countryside. Even now the US's co-op power companies serve three-quarters of the country's huge landmass. These were set up as a result of the Roosevelt New Deal in 1935. This established the Rural Electrification Administration (REA) that responded to the need for work during the Depression and the lack of provision of electrical services to rural areas – only 10 per cent of households in rural areas had power. But it changed the geography of the US, allowing rural areas to diversify from just agriculture to a more varied pattern of economic activity. Traditional utilities were unwilling to invest in the countryside. In July 1935, a group of utility company executives wrote a report in which they claimed that, in light of their earlier extensive research work, 'there are very few farms requiring electricity for major farm operations that are not now served'.[2] By 1953, the REA had managed to reverse rural energy poverty – only 10 per cent of households in rural areas remained without power. The REA (which has since been subsumed into the US Department of Agriculture) originally lent and insured loans for rural electrification projects at a heavily discounted rate. This facility still exists and the rate at the time of writing varies between 0.375 per cent and 4 per cent depending on the term of the loan. This is a useful model for countries such as India – with similar issues of access to energy distribution infrastructure.

At the beginning of the 20th century, people in the US were already sensitized to the exploitation of monopoly by the rail, oil and steel companies. The US invented the system of public utilities commissions to license and regulate the tariffs of energy. In 1907 the New York Public Utilities Commission was issuing reports on tariffs for steam, streetcars, gas and electric utilities.

The Gas Industry

The earliest development of the gas industry took place in the UK. Town gas was made from the incredibly polluting roasting of coal in the absence of air. The Scottish engineer William Murdoch developed the technology in the 1790s. (He is also famous for his work with James Watt on the steam engine.) Town gas was widely used in the UK from the 1820s for street lighting. Local private monopolies sprouted up in cities throughout the UK to provide street lighting for the well-to-do. In Birmingham, where Murdoch spent his adult years, two companies, Birmingham Light and Coke Company (established 1819) and Staffordshire Gas light company (established 1825), sold lighting services mainly to the municipal authorities.

But while investor-owned companies invented and introduced the new technologies, local and socially motivated agents were responsible for their expansion and extension to poorer communities. Manchester street commissioners built the city's gas works in 1817.[3] The most celebrated local government figure Joseph Chamberlain directed Birmingham's municipalization programme. He believed 'all monopolies which are sustained in any way by the state ought to be in the hands of the representatives of the people, by which they are administered, and to whom the profits should go'.[4] It was only through his political manoeuvrings that the city took over the profitable gas monopolies and used the revenues to improve the living conditions in the city and pay for the clearance of slums and provision of water and sewerage services. In the US, local government played a more muted role in direct investment but municipally owned energy companies are found in most states. Within California, Los Angeles Department of Water and Power (LADWP) developed as a wholly owned city business. In Sacramento, the

Sacramento Municipal Utility District was created by the protracted and litigious takeover of investor-owned assets by fiat. The purpose of municipal ownership was to extend access to utilities to areas and communities that the private sector was not so interested in serving, to check the abuse of monopoly or at least to use the revenues from monopoly for social purposes.

It wasn't until the 1950s in North America and 1960s in Europe that gas (natural rather than town gas) became cheap and prevalent enough to be used not just for lighting but also for heating and cooking. A similar story could be told over many cities across Europe with municipal investment in local gas infrastructure to ensure that all citizens had access to night light.

Stages in the Delivery to Our Homes

The delivery of power and gas to our homes and communities isn't just a story of generating electricity or extracting gas. Billions of dollars of investment go into wiring and piping Europe and North America to convey energy to our homes. It's a story of mobilizing vast amounts of capital; it's a story of solving the complex logistic challenge of meeting 24/7 demand; and it's a massive customer service challenge to ensure households and businesses are billed regularly and fairly. Increasingly it has become a public policy challenge. As we saw in the last chapter, the agenda for the gas and electricity industry has shifted from simply supplying energy cheaply to cutting carbon emissions, moving to more secure energy sources and ensuring energy is affordable.

Gas and electricity supply is a major economic activity, creating demands and high-quality skilled employment in other industries. Behind the scenes, financial whiz-kids trade power and fossil fuels contracts moment by moment. Highly trained engineers install gas and electrical appliances and maintain the power and gas lines. Knowledge businesses have evolved to ensure research, planning, contractual niceties and the lobbying of government are taking place to deliver energy to communities. And then there is the vast hinterland of industries involved in providing the capital items and services to the industry. Figure 3.1 shows the different actors that participate in providing gas and power to our communities.

Some aspects of this chain of activities are a natural monopoly. It costs more than $1 million to lay 1 kilometre of pipe or wire. Society sees no sense in two competing sets of infrastructure serving each home. So the conduct of these monopolies has to be acceptable to the general public. In the US, each state has established a public utility commission to set tariffs on the investor-owned companies in areas where the utility has a monopoly right to serve all customers in an area (which is increasingly rare in Europe but still found in most states in the US where the regulator sets the electricity tariff). In countries or states where there is retail competition, the regulator just sets the price of the transmission and distribution services, which the supplier has to pay.

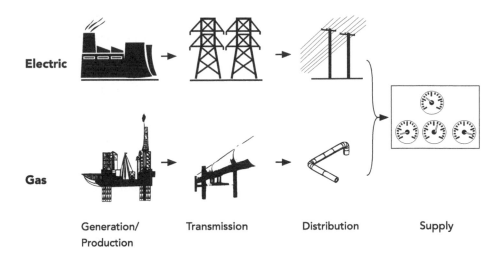

Electric

Gas

Generation/   Transmission   Distribution   Supply
Production

Figure 3.1 The different segments of the electricity and gas industry

Increasingly governments are asking the regulators to set the industry other goals too: standards of customer service and environmental goals. Regulators typically only have duties with regards to the investor private sector energy companies. It is assumed that public and co-op-owned companies can manage without such over-the-shoulders regulations.

When we turn on an electric switch or a gas heater this sets off a commercial ripple that cascades through many different economic actors involved in producing, trading, transmitting, distributing and retailing the energy source.

Electricity is *generated* either by burning fossil fuels or biomass, from nuclear reactions, from the natural flows of wind, water and tide or from sunlight. High voltage lines *transmit* power long distances and low voltage *distribution* lines convey power the last few miles to communities. There are also the *retail* services of reading meters and billing customers. Electricity is expensive to store for any length of time – so in order to provide customers with a reliable and continuous supply there is a vast management system to control the flow of power, information and finances. One agency (often the transmitter) acts as the *system operator* and ensures that the quantity of power on offer matches demand, and that the transmission lines are not being asked to carry more power than they are designed to. Like orchestra conductors, system operators coordinate electricity plants to adjust their output moment by moment to ensure there are no black outs. Within Western Europe there are a number of separate systems that are run independently of one another. The UK has its own Transmission System Operator, Scandinavia has the Nordel network, and Germany and The Netherlands operate within the European Network of Transmission System Operators for Electricity

(ENTSO-E). In the US there are similar regional systems. Traders buy and sell power, gas and coal contracts, and carry out trades to insure their companies against price or currency movements.

Power used to be produced in small local plants but the trend over the last few decades has been to construct large centralized coal and gas power plants, often remote from centres of population (which avoids the blighting of property and ensures compliance with local air quality regulations), close to cooling water and with good access to fuel sources. The largest non-hydro power stations are 4GW or larger in size and include the 6.8GW Bruce nuclear power plant in Canada, the 8GW Kashiwazaki-Kariwa nuclear power plant in Japan and the 4GW Drax coal plant in Yorkshire. Each of these plants can serve the power needs of several million people and necessitates massive investment in transmission and control infrastructure. Once these infrastructure costs are incurred there is an almost automatic impulse to reuse these sites when the power plant comes to the end of its life. But in the case of Drax, both the fuel source and the customers have moved as the UK has replaced its indigenous coal production with imports and as the nation has de-industrialized.

Around 3 trillion cubic metres of natural gas is *extracted* a year. The world's biggest producer (and consumer) is the US, which is responsible for around a fifth of world production. The world's other large producers are Russia and Canada. Gas is internationally traded; countries such as Norway, Algeria and Iran are important exporters of gas. Countries are now making large investments to receive natural gas – the UK's Isle of Grain gas terminal cost almost £800 million and will be capable of supplying around a fifth of UK's gas needs. Gas is *transmitted* through gas pipes and stored in above-ground and below-ground chambers to buffer the fluctuation in demand with supply. The US has 305,000 miles of transmission pipes, 1400 compressor stations that pump the fluid, 24 hubs to interconnect different pipelines and 400 natural underground stores. Gas is then *distributed* to homes and communities through a local gas distribution system. Energy companies *retail* gas, metering and billing customers for the gas they use.

Distributed Generation

The trend has been for power plants to become larger and for gas production/import to become more centralized. This allows firms to exploit economies of scale in production, trading and long-distance transmission. But there is an alternative way of supplying energy and energy services that is smaller in scale and better fitted around local needs: distributed energy systems tap smaller, localized sources of energy such wind, hydro, biomass or energy from waste, while energy efficiency, by which we mean investment or changes in the way buildings or appliances are built and run, reduces energy demand.

Heat and power plants can be installed in or near the communities they serve. This reduces the costs and nuisance of constructing transmission lines.

Transmission into major centres of population can act as a bottleneck in meeting a community's new energy needs. Toronto only has two transmission lines feeding it; both are operating close to their capacity, and the political resistance to constructing a new line of pylons is substantial. More importantly, producing electricity near to where it is consumed means that heat wasted from conventional power stations can be used to keep homes and offices warm. A new combined-cycle gas turbine power station only utilizes half the energy in the fuel, a coal-fired power station uses a maximum of 40 per cent, while a nuclear power station uses less than this. A well-designed CHP plant can recover 80 per cent of the energy from the fuel, resulting in at least a 30 per cent improvement in efficiency than the separate use of gas for heat and power.

Local production of energy might alter the psychology of energy use. Why should anyone nowadays worry about conserving energy if someone far away has to put up with the nuisance of generation and transmission? But if new demand has to be met by plant close to where you live, which you are likely to encounter every day, then it might change your attitude to conservation and to objecting to power production.

Denmark has been at the forefront of distributed generation, making extensive use of cogeneration and DH to provide hot water to homes, and introducing small community-owned wind farms.

Energy Efficiency

There's a hilarious scene in the film *Finding Nemo* where a group of sharks are chanting the mantra 'Fish are friends' in a slightly delusional bout of auto-hypnosis. Even Prashant's son – who was just three years old at the time – picked up the absurdity of the situation: 'That's silly, dad', he said in his most non-plussed voice. Absurd it was. Which dumb schmuck would stick his head in a 'vegan' shark's mouth?

Well, welcome to energy efficiency and renewables delivery policy on both sides of the Atlantic. Investor-owned companies exist to maximize the returns to their shareholders – which means selling as much energy as possible. Yet governments are mandating energy companies to achieve prescriptive targets to install energy efficiency measures to conserve gas and electricity. These programmes use customer money to fund the purchase of appliances and insulation to cut the demand for the utility's product. Renewables policy similarly requires utilities to buy or generate electricity that they wouldn't choose to through market forces – since it wouldn't meet their investment criteria.

Energy is well suited to commoditization; energy efficiency is not. Instead of being purchased in kWh or barrels, energy efficiency is embedded into the design of products, or comes about through deliberate building practices such as draughtproofing or loft insulation, or it is a learnt new behaviour. While an electron is an electron, and a molecule of methane a molecule of methane, the energy efficiency of a product or practice is context specific. Its efficacy depends on the system in which it is embedded. An energy-efficient light bulb is best deployed in a room illuminated many hours a day; it is wasted if is kept in a drawer.

Energy efficiency is also a surprisingly fertile ground for innovation. For instance the pursuit of energy-efficient window design has led to the replacement of air as a filler gas with argon, then xenon and krypton. Now manufacturers are experimenting with panes that hold the ultimate insulator – a vacuum. The glass pane too has been modified and is itself a chemical matrix doped to changed its reflectivity and emissivity to light. New better-insulating materials are now used to make the spacer bar that separates the two glass panes.

The Businesses Supplying Energy in Europe

In northwest Europe energy provision is concentrated in just a handful of firms. These companies own businesses spanning the generation of power or the production of gas through to its supply. (Though the ownership of the wires and pipes has to be separately accounted for to ensure that the competitive business is not subsidized by their monopoly businesses.) The biggest companies generate power and sell it directly to their supply businesses. This is true for gas but to a lesser extent than electricity.

The largest European energy companies used to be national public entities, but most EU countries have privatized them. Germany's RWE and E.ON were privatized several decades ago, as were energy companies in Spain (part of post-Franco liberalization). The UK's publicly owned Central Electricity Generating Board (CEGB) was privatized in the 1990s and split up. In England and Wales the generating parts of the CEGB were split into National Power (subsequently renamed npower) and Powergen. (Nuclear power stations remained in public ownership until 1996.) Twelve regional electricity companies were created to handle the distribution. In Scotland, Scottish Power and Scottish Hydro Electric were created. These continued to carry out generation and distribution, as well as supply.

There have been exceptions to the privatization trend. Vattenfall is still wholly owned by the Swedish state (although the government floated the idea of part privatization in August 2010). EDF is 85 per cent owned and GDF Suez is 36 per cent owned by the French government; Dong is 75 per cent owned by the Danish state. This makes these companies susceptible to national interest issues such as employment creation, sourcing of components and energy security issues, which are often in conflict with the Single European Market, which treats issues of sovereignty and national determination as subordinate to the desire to extend the pan-European market.

The EU's Single Market Programme has promoted competition in energy markets, but as ever with EU initiatives, different member states have responded in very different ways. Britain has the most competitive energy market of any EU country (Northern Ireland has a very different energy market). The energy companies in Britain no longer enjoy monopoly rights to sell electricity or gas to customers within a defined area. Sweden has the second most competitive, where there is a lot of competition in electricity but little in gas. Finland, The Netherlands and Denmark also have competitive electricity markets but not much competition in the gas market. Spain has the second most competitive gas market (after the UK) but a less competitive electricity market.

The German energy market is significantly less competitive than the UK market. The German government has been opposed to ownership unbundling of networks, so integrated companies still carry out generation, distribution and supply. Four companies dominate electricity (E.ON, RWE, Vattenfall, EnBW), and these four plus EWE dominate gas.

There is much less competition in French energy markets. A comprehensive analysis of EU and G8 energy markets carried out by Oxera did not include analysis of France because it was thought to have insufficient competition to clear the initial hurdle.[5]

The increase in competition in energy markets has led to pan-European energy companies, as Table 3.1 shows. In the UK, npower was bought by RWE, and Powergen by E.ON (so is now called E.ON-UK) (see Box 3.1). Iberdrola bought Scottish Power in 2006, and is also active in Germany, France and Italy. EDF is active in the UK and The Netherlands. Vattenfall is active in The Netherlands, Germany and the UK.

The pan-European identity of many firms means that billion euro decisions about investment in new plant, which can impact on a country's energy plans for a whole generation, are made in remote corporate headquarters in Dusseldorf, Essen and Paris and are not so susceptible to arm twisting by another country's

Table 3.1 Concentration and international reach of the European energy companies (electricity, gas and heat sales)

	Home country	Other markets	Revenue 2008 (€ billions)
E.ON	Germany	UK, Sweden, US	82
EDF	France	UK, Germany, Netherlands, Sweden Austria, US	66
RWE	Germany	UK, Netherlands,	48
GDF Suez*	France		27
Centrica	UK	US	26
Vattenfall	Sweden	Germany, UK, Netherlands, Denmark, Finland	20
Dong Energy	Denmark	Norway, UK, Netherlands	5

Note: * GDF Suez revenue data refer to 2007, electricity figures refer to generation *not supply.*

Source: Corporate annual reports and GDF Suez (2009) 'Reference document', www.gdfsuez.com/document/?f=files/en/ddr-2009-en.pdf

---

Box 3.1 E.ON – the largest energy utility in the Western world

E.ON is the largest investor-owned energy supplier in Europe. Like many companies it was created as a public utility and has transitioned to a monolithic private company. It serves 30 million customers in 30 countries and is particularly dominant in Germany, the UK and Sweden. Its sales in 2008 were €87 billion. It was formed from the merger in 2000 of two German conglomerates – VEBA and VIAG – which were founded by the German government in the 1920s to develop the country's industrial base.

The company's 2010 'Energy Fit' TV advert shows a clean-cut family using an energy monitor for the first time. But the corporate strategy is more hard-nosed. It focuses on principles that include having a presence in every part of the energy chain from the production of gas, the generation of electricity, the transmission of energy through to the local distribution and sales. The strategy is premised on growth – outward into new markets through the increasingly liberalized and interconnected European markets and also deeper with greater and greater market share. The business model is not really that different to the EU's other energy behemoths: Germany's RWE, France's EDF and GDF Suez (formed from the merger between the French gas and water utilities), Sweden's Vattenfall, UK's Centrica, Spain's Iberdrola and Denmark's Dong.

---

government. Instead, countries have to dazzle companies' finance officers and shareholders to secure inward investment for these essential services. This is a major undertaking and it is difficult to reverse change in a country's capacity to determine its energy mix.

Despite the promotion of competition, European energy policy has not created diverse energy markets. Large companies dominate. For example, in the UK there are just six major suppliers and virtually no hinterland of mid-sized competitors. Apart from these huge companies there are a handful of small specialist companies serving environmental or business customers, which go bust from time to time. The US energy market is very different.

The Businesses Supplying Energy in the US

In the US there is hyper-fragmentation of the industry. The comedy show *The Simpsons* isn't far off the mark in its depiction of the mom-and-pop nuclear power-based utility owned by Montgomery Burns. Altogether the country has 3000 energy utilities, each licensed and regulated at the state level. Some integrate generation with distribution and supply, while others just distribute and supply electricity. Tiny Vermont with a population of just 600,000 has 20 different energy suppliers. Over two-thirds of the US's energy suppliers, are publicly owned companies serving a city or some other political boundary. Only 240 of the 3000 are private, investor-owned companies.

Table 3.2 *Selected electric utility data by ownership, 2000*

	Investor-owned	Publicly owned	Co-operative	Federal	Total
Number	240	2009	894	9	3152
(%)	7.6	63.7	28.4	0.3	100
Revenues ($ billions)	170	33	21	1.2	224
Revenues (%)	75.6	14.7	9.1	0.6	100
Sales (million MWh)	2438	517	306	49	3310
Sales (%)	73.7	15.6	9.2	1.5	100

Note: Data are based on calendar year submissions.
Source: Energy Information Administration, Form EIA-861, 'Annual Electric Utility Report'.

Table 3.3 *Revenue and geographic reach of US power companies*

	Headquarters	Other markets	Electricity sales 2008 ($ billions)
Investor-owned			
Southern Company	Atlanta, GA	AL, MI	16
American Electric Power	Columbus, OH	TX, VA, TN, IN, KY, OK, LA	13.5
Duke Energy	Charlotte, NC	SC, OH, KY, IN	12.7
Edison International	Rosemead, CA		12.3
Municipally owned (2008 data)			
Salt River Project	Phoenix, AZ		2.9
LADWP	Los Angeles, CA		2.7
New York Power Authority	White Plains, NY		2.7
Santee Cooper	Moncks Corner, SC		1.6
Co-operative			
Southern Maryland Electric Cooperative	Hughesville, MD		0.45
Vermont Electric Cooperative	Johnson, VT		0.07

Note: GA – Georgia, AL – Alabama, MI – Michigan, OH – Ohio, TX – Texas, VA – Virginia, TN – Tennessee, IN – Indiana, KY – Kentucky, OK – Oklahoma, LA – Louisiana, NC – North Carolina, SC – South Carolina, IN – Indiana, AZ – Arizona, CA – California, NY – New York, MD – Maryland, VT – Vermont.

Source: Various annual reports

But these numbers belie the true importance of private, investor-owned companies. In 2000, the 240 private companies sold three-quarters of the electricity produced. In contrast, the publicly owned companies took just 15 per cent of revenue. The remaining 10 per cent was taken by 894 rural electric co-ops that were set up in the 1930s to distribute electricity to farming communities.[6] There are also nine federal energy utilities such as the Tennessee Valley Authority and those operated by the Army Corps of Engineers (see Table 3.2).

Some of the private companies operate in many states. For example, American Electric Power, based in Ohio, is also active in Texas, Kentucky, Michigan and several other states (see Table 3.3).

The largest private US energy companies are much bigger than municipal or co-op companies, but still small by European standards. E.ON's revenues are five times those of the largest US utility, Southern Company.

Why Private Companies Aren't Best for Energy Efficiency Work

If you want every i dotted and every t crossed in your energy efficiency programme you will get just that. But why would any utility go beyond the bare minimum of complying with your order?

Utility CEO responding to a regulator 'ordering' the utility
to install energy efficiency[7]

The existing business model of most energy companies is to 'predict and provide' – anticipate what the level of demand for energy will be and then supply enough energy. The more energy they supply, the more profit they will make. This makes them less than wholeheartedly supportive of energy efficiency. Politicians and regulators have tried to overcome this by 'decoupling' profits from the amount of energy sold, as the next chapter explains. This has had some impact but has not altered the existing business model of energy companies in which selling more is regarded as the core business objective.

The above quote is taken from Peter Fox-Penner's illuminating book *Smart Power*.[8] Many industry insiders are conscious, some even queasy, that the delivery of energy efficiency depends on making firms carry out commercial hari-kari. In the US, regulators in many states require utilities to use ratepayer money to deliver energy efficiency programmes. In the UK, each energy supplier is set targets by government to install an amount of energy efficiency in people's homes. Some provinces in Canada are asking utilities to do the same. In Denmark, utilities have also been mandated to improve energy efficiency.[9] Substantial resources are going into these energy efficiency programmes – in the UK some £1 billion a year is spent on energy efficiency measures, which is around 3 per cent of the bill.[10] Regulators in the US typically ask utilities to spend up to 2 per cent on their energy-saving programmes. These programmes have been successful in reducing energy use. They have encouraged the roll-out of energy efficiency commodity products such as light bulbs and white goods. But have we stopped to ask whether these are the correct institutions to deliver energy efficiency? Billions of euros and dollars of investment in people's homes are being made by the energy companies.

The UK's Consumer Focus undertook a survey of consumer confidence in 45 different markets: gas and electricity came bottom in 2009 alongside the reviled pension product market.[11] In Norway, in a similar survey, electricity came 30th out of 41 industries.[12] The European Commission publishes a Consumer Markets Scoreboard across the 27 EU member states. Of 19 services and goods markets, the gas and electricity markets were the fifth and sixth lowest scoring, beating only public transport, fixed telephone and postal services.[13] Consumers trust energy companies to the extent they do not expect to be mugged by their staff, but experience shows that where retail choice exists, the sales staff employed by utilities mislead customers into buying unsuitable tariffs, and the quality of service is often woeful.[14]

Improving the energy efficiency of existing buildings, one of many communities' biggest challenges, doesn't even play to the strengths of utilities. Unlike appliances and light bulbs, home retrofits, especially of older homes, require the contractor to tailor the service to individual homes, have sensitivity to community-level politics to navigate aesthetic mores, and have excellent inspection and verification procedures. The energy companies are not the first port of call when people need minor construction work or maintenance, but such building repairs and redecoration will usually be the best triggers for installing insulation.

Because building projects are not utilities' core strengths they usually outsource the delivery of the programmes to local firms. In practice the utility does little more than market services in line with their analysts' determination of the cheapest means of complying with the regulations. As a result they have only a limited interest in ensuring the work has been carried out well, or was even merited – energy companies spend less than 1 per cent of their energy efficiency budget on evaluating the effectiveness. When the authors asked the UK energy regulator whether it or the firms verify the schemes, to see if the energy efficiency measures were properly installed and operating as intended, it said: 'The suppliers are required to check the quality of their contractor's installations, but not do any thermal monitoring as to do this for a representative sample would be expensive'.[15] Results on how much consumers will save are modelled by the companies, rather than measured, and the regulator does not check.

Little work has been published on evaluating how well the work is carried out. The majority of savings is presently attributed to cavity wall insulation and loft insulation but the amount of energy saved depends on how diligently these are fitted, and how well the builder makes good his work. One study by the Energy Saving Trust in 2008 looked at 70 homes that had cavity wall insulation fitted under the UK government's Warm Front programme (rather than the energy company schemes).[16] This suggests that the coverage of the insulation was 40 per cent lower than planned because installers had failed to fill in the cavity fully. Another study of 1500 homes covered by Warm Front that measured the internal temperature and humidity of homes and looked at the heating bills concluded that: 'the potential improvement in energy efficiency from the installation of draught stripping, insulation and a gas central heating system was not observed, and there appears to have been no reduction in fuel consumption as a result of the Warm Front measures'.[17] This was in part because people heated their homes to a more comfortable temperature but it was also due to revealed weaknesses in the modelling (which underestimates the loss of heat from ventilation) and poor workmanship (for instance failing to seal gaps and holes in the wall when boiler exhaust pipes are fitted).

We don't know how well insulation installed by utilities performs as they do not publish any results. In the UK, current spending is around £1 billion a year and yet there is little evaluation of performance, and none in the public domain. In the UK, utilities spend around £200 a home on cavity wall insulation (customers contribute around the same amount). Utilities are interested in discharging their obligations as cheaply as possible; on average installation is completed in just two hours. In the US, weatherization undertaken by the Community Action Committees takes as much as two days, as they carry out a more complete treatment of the home and better checks on quality. The cost is of course much higher.

Neither of these business models ensures that the firms actually benefit from installing energy efficiency well. To do this the business would need to be exposed to the real performance of the home, and not just a computer simulation of performance. The business model that delivers this is the ESCO, where the customer pays the utility for an energy service (keeping the home at a particular temperature for a year), rather than for how much energy is used. This is a radical departure from the present system and has its own problems. How do we retain the incentive for the customer to use energy sparingly?

Changing the business model will not happen of its own accord. It requires the development of firms that provide customers with energy services (a warm home, lighting and the bill for the provision of these services), rather than the kWh of energy used by people. The ESCO business model means firms have an incentive to invest in energy efficiency. Firms such as Dalkia provide energy services, especially to offices, but they have not prospered, in part because they are more exposed to fossil fuel price risks than their vertically integrated competitors.

The contrast between the approach of investor-owned utilities and the rural electric co-operatives in the US should be noted: 92 per cent of the members of the National Rural Electric Cooperative Association actively educate their consumers (who are members and owners) about energy conservation. However, some voices in investor-owned companies do accept the need for significant change. British Gas announced in September 2010 that it would invest £30 million in installing energy efficiency measures in homes at no upfront cost, with repayments made by savings in the customers' energy bills. The head of Centrica (of which British Gas is part), Sam Laidlaw, made a powerful speech at the time calling for the energy industry to have 'a fundamental change of mindset'. He accepted that 'we are not there yet', but the speech suggests that personally he is there. He went on to say that the scale of the challenge:

> means a company like British Gas must be as much about energy saving as it is about energy supply. Decarbonising our power supply will get us a long way towards cutting carbon emissions, but it is not enough. We need to look closer to home. Britain's households are responsible for over a quarter of the UK's CO_2 emissions. Cleaner electricity can go some way towards cutting this. But it won't make a difference to the gas we use to heat our homes and cook our meals. We are already addressing domestic emissions by improving the energy efficiency of new homes, with sharply improved building standards. But we can't ignore our existing housing stock. Two thirds of homes today will still be lived in by 2050. For most of us, today's home will also be the home of the future. Yet their energy efficiency is the worst in northern Europe. There are still over 4 million inefficient boilers in the UK. Six million homes are without cavity wall insulation. Seven million without proper loft insulation. Just putting this right could cut the carbon footprint of the average British home by around 20 per cent.[18]

This was only a speech, of course, and words are easy (and cheap). Nevertheless, British Gas is doing quite well on energy efficiency. This suggests that large investor-owned utilities shouldn't be written out of the story altogether. They are needed for the enormous investments in offshore wind farms and other low-carbon electricity supply options. They could also play a role in energy efficiency. But new actors in energy efficiency are urgently needed.

The desire to maximize profit is the main reason why private companies aren't ideal for energy efficiency work, but not the only one. Another reason is that they often have to borrow at higher interest rates than public bodies such as local governments. When large amounts of capital are needed, as is the case for energy efficiency retrofits and renewables developments, the project owner usually finances the project from borrowing even if the project initiators have substantial funds in hand. The project owner has to find willing lenders and then to negotiate reasonable loan terms. Local government can get lower interest rates than private companies because the risks are lower. As we show in Chapter 6, in continental Europe loans are often backed by local or central government, reducing the cost of capital.

Why Aren't Energy Efficiency Businesses Doing Better?

Given these drawbacks of existing energy companies, why aren't new energy efficiency companies doing better? Energy efficiency businesses are in a sense oxymorons, defined by what they don't do – unnecessarily use energy. Energy efficiency firms are also highly disparate and fragmented. The membership of the European Alliance of Companies for Energy Efficiency in Buildings (EuroACE) includes window manufacturers, heating control firms, the conglomerate Honeywell and building materials providers. The shop front for energy efficiency is usually provided by the notoriously unreliable building sector. The provision of decentralized energy is through a handful of small specialist renewable manufacturers and installers, waste companies such as Veolia and subsidiaries of the big energy utilities.

Even the largest of firms selling energy efficiency products are small compared to the energy suppliers: the Danish windows company Velux had total sales of only €2.1 billion in 2009. The German building materials manufacturer Knauf sold goods to the value of €5.6 billion in 2009 (less than 10 per cent of the value of E.ON's sales). Others are hugely diverse enterprises such as Honeywell or Phillips that sell energy controls and compact fluorescent light bulbs along with so many other things. Often they are diversified and might not consider energy efficiency as an output. Honeywell sells controls and systems for energy conservation, Danish Danfoss sells valves and controls. Some companies exclusively supply energy-efficient apparatus, while Megaman specializes in inventing and selling energy-saving lighting.

Aside from the mainstream energy industry is the state-of-art obsessive element. There are many approaches to ultra-low energy building: the best known are the Passivhaus from Germany, zero emissions developments (ZED) in the UK and the net zero movement in the US. These demonstrate what is possible if advanced building science ideas are applied well. But we are a long way from deploying these in new buildings and only some of these techniques can be retrofitted to old homes.

So those involved in energy efficiency need to accept some of the blame for lack of success. However, the main reason why energy efficiency businesses have not been more successful is lack of demand. Due to low energy prices, what demand exists is largely a result of government policies. Tougher energy performance standards for new buildings have become commonplace since the 1970s, but have no impact on the massive stock of existing homes. There is little spontaneous demand for consumers to install measures to improve the energy efficiency of their homes. In some countries – Scandinavia and The Netherlands, for example – energy taxes have increased consumer demand for energy efficiency. These countries' governments have recycled some of the tax revenues as grants for energy efficiency and renewables. Germany has used low-interest loans to promote energy efficiency and renewables. Energy and carbon taxes, and grants for energy efficiency and renewables, are covered in Chapter 7. Other policies, and the politics that have led to them, are the subject of the next chapter. But it's worth noting here that the fragmentation means that energy efficiency companies make a weak lobby. UK ministers quite regularly capitulate to pressure from the building industry and withdraw proposals to mandate tough retrofit energy efficiency standards when homes are extended or significantly adapted.[19] The established firms in construction and energy supply still have the ear of policy-makers. There are some huge challenges ahead of us still.

Notes

1   DECC (2010) 'Press release for Spalding gas-fired power station', 11 November, DECC Press Release: 2010/118.
2   National Rural Electric Cooperative Association, www.nreca.coop/members/History/Pages/default.aspx.
3   Hunt, T. (2004) *Building Jerusalem – The Rise and Fall of the Victorian City*, Weidenfeld & Nicolson, London.
4   Hunt, T. (2004) *Building Jerusalem – The Rise and Fall of the Victorian City*, Weidenfeld & Nicolson, London.
5   Oxera (2007) 'Energy market competition in the EU and G8, preliminary 2006 rankings', http://webarchive.nationalarchives.gov.uk/+/http://www.berr.gov.uk/files/file44272.pdf.
6   US Energy Information Administration (2009) 'Summary statistics for the US', www.eia.doe.gov/cneaf/electricity/epa/epates.html.
7   Cited in Fox-Penner, P. (2010) *Smart Power*, Island Press, Washington, DC.
8   Fox-Penner, P. (2010) *Smart Power*, Island Press, Washington, DC.
9   Hamilton, B. (2010) 'A comparison of energy efficiency programmes for existing homes in eleven countries – Report to Decc', Regulatory Assistance Project, www.raponline.org/docs/RAP_Hamilton_ComparisonOfEEProgrammesForExistingHomesInEleven Countries_2010_02_19.pdf.
10   Lees, E. (2009) 'Evaluation of EEC2 and initial thoughts on CERT', presentation to Ofgem, www.ofgem.gov.uk/Sustainability/Environment/Policy/EnvAdvGrp/Documents1/Eoin%20Lees%20EEC2%20evaluation%20presentation.pdf .
11   MORI (2009) *Report on the 2009 Consumer Conditions Survey Market*, research survey conducted for Consumer Focus, MORI.
12   Berg, L. (2010) 'The Norwegian Consumer Satisfaction Index', SIFO, The Consumer Ombudsman.
13   DG SANCO (2009) *The Consumer Markets Scoreboard*, 2nd edition, Office for Official Publications of the European Union, Luxembourg.
14   Ofgem (2008) 'Energy supply probe', www.ofgem.gov.uk/Markets/RetMkts/ensuppro/Documents1/Energy%20Supply%20Probe%20-%20Initial%20Findings%20Report.pdf, Accessed 29 August 2010. See Table 4.1. In this table only 52 per cent of gas customers and 58 per cent of electricity customers that switch tariffs as a result of direct sales benefit from the change. Almost half of customers would have been better off staying with their original supplier had the salesperson not persuaded them to switch!
15   Personal communication to Prashant from an Ofgem official, 2010.

16 Energy Saving Trust (2008) 'Thermal transmittance of walls of dwellings before and after application of cavity wall insulation', Report number 222077, www.decc.gov.uk/assets/decc/what%20we%20do/supporting%20consumers/sustainable%20energy%20research%20analysis/1_20090710110441_e_@@_8thermaltransmittanceofwallsofdwellingsbeforeandafterapplicationofcavitywallinsulation.pdf.

17 Hong, S., Oreszczyn, T. and Ridley, I. (2006) 'The impact of energy efficient refurbishment on the space heating fuel consumption in English dwellings', *Energy and Buildings*, vol 38, no 10, pp1171–1181.

18 Speech by Sam Laidlaw, 16 September 2010, Royal Society for the Encouragement of Arts, Manufactures and Commerce, www.centrica.com/files/pdf/Sam_Laidlaw_RSA_Speech_16_Sept.pdf.

19 Warren, A. (2010) 'Britain pays a high price for personal prejudice', UK Association for the Conservation of Energy, www.ukace.org/publications/ACE%20Warren%20Report%20(2010-07-08)%20-%20Britain%20pays%20a%20high%20price%20for%20personal%20prejudice.pdf.

Chapter 4

Why Governments Aren't Delivering

Democracy however desirable is not an effective device for solving global or transnational problems.

Eric Hobsbawm (1995)[1]

Energy is too important to leave to the market. A free market would either deliver the cheapest energy, however insecure or polluting this was, or not deliver sufficient energy. Energy is a central economic, social and environmental issue. It is therefore also a central political issue. All governments, even those that stress their free-market credentials, have an energy policy, or more usually several energy policies: taxes, cap-and-trade, subsidies, regulation, land-use planning. Some governments have used all of these. But no government or country is doing well enough to secure the energy transformation that we all need. Should we therefore conclude that the global and transnational problems of climate change and energy security are, as Hobsbawm suggests, too important to leave to democracy?

That is not our view. We believe that democracy is a central part of human progress. But the performance of democratic politicians clearly needs to improve substantially. Short-termism and love of soundbites and photo shoots must be overcome. The endless setting of targets and timetables and reorganization of bits of government must end, with the focus shifting to regulation and delivery. The power of special interest groups must be overcome. Politicians must stop defending their own patch and accept that powers should pass to the tier of government best able to achieve the transformation.

Can this happen? We think that it can, but it won't be easy. This chapter looks at the open and hidden objectives that politicians have for energy policy. The chapter then summarizes the role and powers of different tiers of government in different countries. This is important because the powers of particular tiers are central to the politics and policies of all those involved. One of the main blockages to sensible policies from local councils in the UK is not political but constitutional. Local government lacks many of the powers held by sub-central government in the US and in other European countries. The word 'federal' is a term of abuse in the UK, flung at those who want European countries to work more closely together – often by those who claim to be very pro-American but seem never to have read Jefferson's *Federalist Papers*. The US and Germany are much less centralized because they have federal constitutions. Scandinavian countries are even more decentralized, giving local government great freedom of action. The chapter then looks at the politics of energy policy in Europe and the US, and ends by seeking to answer the question: why aren't politicians delivering more?

Declared Objectives of Energy Policy

There are six declared and uncontroversial objectives to energy policy: economic development, keeping the lights on, energy security, energy efficiency, afford-ability and employment support. These objectives are shared by almost all politi-cians, for the reasons discussed in Chapter 2.

The objectives on climate protection is selectively declared. The need for climate protection is not widely agreed in North America, while it is in most European countries. However, the appropriate response certainly isn't agreed within Europe. Politicians often underplay climate issues and stress economic or energy security issues in order to minimize political conflict.

Energy has always been seen as an economic issue, and its affordability as a social issue. From the early 20th century it has also been recognized as an environmental issue. Air pollution is a major problem in urban areas. At first this came mainly from burning coal for heating homes. The 1956 UK Conservative government passed the Clean Air Act, which banned the burning of coal in UK cities and banished the infamous London smogs. The primary cause of air pollu-tion in urban areas used to be coal but is now more often oil. The spread of the car may have done wonders for human society and happiness (at least for those able to afford one), but it brought major pollution issues. California led the way in reducing this problem, not by telling people not to use automobiles (politically difficult anywhere, but particularly so in the US) but by using the power of govern-ment to require a technological solution. Catalytic converters, which cut about nine-tenths of the toxic pollution coming out of tailpipes, have been mandatory on new cars in California since the 1970s. The rest of the US followed California, as did, eventually, the EU, where catalytic converters have been mandatory since the early 1990s.

Some of the pollutants coming from vehicles are sulphur dioxide and nitro-gen dioxide. These damage human lungs, but are also a major cause of acid rain, which destroys forests and wildlife. This issue put the focus back on coal, since coal-fired power stations are major sources of acid rain gases. In the1980s there were widespread campaigns against acid rain in the US and Europe, leading to the cap-and-trade systems on sulphur and nitrogen in parts of the US, and regulations on power stations in the EU.

By the mid-1980s, the dominant energy issue of concern to 'greens' was climate change. Since then, European politicians have been active in talking about the need to reduce greenhouse gas emissions (though less active in delivering actual reductions). So the need to control climate change should be added to the list of declared objectives, for European politicians at least. It is also an open objective for most US Democrat politicians, including President Obama and Energy Secretary Chu. All politicians who wish to reduce greenhouse gas emissions are up-front about this.

But, as discussed in Chapter 2, climate change is not uncontroversial: certainly not in the US and also not in the UK. This has led some politicians to stress other reasons for developing alternative sources of energy. The UK Conservative Party's energy policy document in March 2010, shortly before the general election, stressed the need to use energy more efficiently and to maximize the use of renewable energy. Both would be excellent ways to reduce carbon emissions. But the document barely mentioned climate change. Instead, it stressed the energy security advantages. This isn't because they aren't interested. While he was director of Greenpeace UK, Stephen went to a meeting with David Cameron, two days after he'd been elected Conservative Party leader. This was a meeting for the heads of UK green NGOs, and Stephen and the others were impressed with Cameron's knowledge and personal commitment. But many in the Conservative Party are 'climate sceptics', so it was sensible politics to play down climate change and play up energy security just before an election.

Even in countries where climate change is not a politically or scientifically controversial issue, politicians realize that the other advantages of sensible energy policies are worth emphasizing. For example, in Germany the employment potential of renewables has been stressed. In Denmark and Germany politicians have also stressed that a switch to renewable energy has the potential for common and community ownership. About 100,000 people own shares in wind farms in Denmark, where 80 per cent of older wind farms began as co-operatives, although more modern ones are more often privately owned or shared. Copenhagen harbour's offshore wind farm has 8000 investors owning 50 per cent, with a local utility owning the rest.

Undeclared Energy Policy Objectives

There are also at least three undeclared objectives to energy policy:

1   Reducing trade union power. Some governments, mainly but not exclusively right-wing ones, aim to break or reduce the power of trade unions. Margaret Thatcher was certainly determined to reduce the influence of the National

Union of Mineworkers and its belligerent leader Arthur Scargill. Angela Merkel's agreement to allow Germany's nuclear power stations to remain open longer than previously planned was seen by some opponents as a desire not to play into the hands of the coal industry, including the trade unions.

2   Developing nuclear weapons. Support for nuclear power has very often been associated with a desire to develop nuclear weapons. There is no clear energy reason why Iran, which has the second largest known gas reserves in the world (after Russia) and the third largest known oil reserves (after Saudi Arabia and Canada), needs nuclear power. There might be a climate rationale for building nuclear, but this is unlikely to be a major selling point for the Iranian regime. Economics don't explain its action either. Nuclear power is low carbon but very expensive. The likely real reason is because it wants nuclear weapons. Since the US made the first atom bomb, every country that has acquired nuclear weapons has done so by developing civil nuclear power first (except Israel, which bought bombs from the French). However, there are many countries with nuclear power that have no nuclear weapons. Merkel's shift on nuclear was not an indication that Germany wants nuclear weapons, and Japan has many nuclear power stations but no nuclear weapons. So not all politicians who support nuclear power can be tarred with the bomb brush.

3   Grandstanding. Politicians like to claim the mantle of being the most committed to 'saving the planet'. This leads to competition between political parties (and in some cases even within political parties), but also competition between countries, states and local areas.

Many of the declared and hidden objectives are overlapping. Merkel's proposal to allow nuclear power stations to remain open longer than planned (which she abandoned after the accident at Japan's Fukushima I Nuclear Power Plant in March 2011) would have reduced Germany's climate emissions and also reduced the power of coal trade unions. The objectives are also in some cases contradictory. Developing renewables is good for both energy security and climate protection, but not (yet) good for affordability.

Powers of Different Tiers of Government

Political discussion and objectives are, of course, heavily influenced by what a particular tier of government can do in constitutional terms. The president of the US is often regarded as the most powerful man in the world but he has little influence over schools in Washington or the proportion of energy that has to be supplied by renewables across the Potomac in Virginia – a degree of powerlessness many European leaders would find surprising. In the UK, the mayor of London has virtually no powers to raise finance (aside from a congestion charge and a precept on local property taxes), no control over individual planning decisions, energy or water tariffs and education, and limited powers over policing and transport. Constitutions and conventions restrain the power any single politician may wield.

The US federal government's key energy and climate powers concern the regulation of interstate commerce and the regulation of pollution from power stations. The federal government uses its interstate commerce role to set minimum efficiency standards for energy-using products, and to set prices for the transmission of electricity and transport of gas across state lines. Beyond this, it can provide funds and incentives for innovations in energy efficiency and renewable energy, but cannot mandate any specific programmes.

It is the state governments that set most energy policy. Each state has a public utilities regulatory commission (PURC) to regulate private sector energy companies and set their energy tariffs. US states implement many types of energy policy, including: setting building standards; requiring energy suppliers to contribute to demand-management efforts; requiring that a certain proportion of electricity comes from renewables; restricting the emissions from a power station through an emissions performance standard; and establishing cap-and-trade systems for CO_2.

Beneath state government is local government. There is no uniform role for local government in the US. Each of the 50 states has its own constitution, so local government in each state has different roles. US localities rely on own-generated revenues for over 80 per cent of their budgets. Their practical power is, however, limited by the extremely small scale of the counties and municipal units in most states. It is not uncommon to have counties with fewer than 25,000 residents, and many municipalities within those counties. County boundaries date to an era when the jurisdictions were defined by how far a farmer on horseback could travel in a day.

Constitutional issues are even more complex and politically fraught in Europe: 'Brussels has greater potential power to shape the energy market design of its member states than Washington has over US states' (Buchan, 2009).[2]

Energy has always been a central part of the 'European project'. The European Coal and Steel Community, established in 1951, was a precursor to today's EU. Since 1986 energy has been part of the Single Market Programme – the intention to remove all barriers to the whole of Europe acting as one market, which is broadly similar to the US interstate commerce approach.

The Commission is the executive branch of EU government. It is supposed to implement agreed policies, and also be the referee between national governments. Commissioners are not elected but selected by national governments. Nevertheless, the Commission is the only institution that can formally propose legislation or regulation. Proposals then have to be agreed, amended or rejected by the European Parliament and national governments, operating in a body called the Council of Ministers. The Council used to require unanimous agreement to any proposal, so any country – even tiny Luxembourg – could block progress. But in 1986 the Single European Act was agreed. This aimed to create a single market. Among other changes, the Act introduced qualified majority voting – qualified in that the votes of big countries are worth more than those of small countries. Countries gave up the national veto. Many energy policy proposals are now dealt with through qualified majority voting, though anything involving taxation still requires unanimity.

The constitutional structures of the member states are very varied. The UK has no written constitution and so has a patchwork of constitutional arrangements. Labour introduced devolution to Scotland and Wales and restored it to Northern Ireland. It also tried to establish elected regional assemblies in England, but this proposal failed in a referendum in northeast England. The Scottish and Northern Irish governments have control of energy regulation and can award planning permission to power stations or wind farms larger than 50MW. The Welsh government would like these powers but has not been given them. The UK government gives planning permission for any power plant over 50MW in England or Wales. The main policy to promote renewables, the Renewables Obligation, was established by the UK government, though it only covers England and Wales. The Scottish and Northern Irish governments have a similar system, though the Scottish system is more generous for wave and tidal technologies.

Much of the country has two tiers of local government: counties and districts. In urban areas, and some rural areas, there is one tier of local goverment. The UK government controls local government by controlling the money – around 80 per cent of local government funds come from the national government. The one area where UK local government does have significant powers is land-use planning. This is understandable and welcome: Britain is a small crowded island with some wonderful scenery. But it also explains why the UK – a windy island – derives so little of its energy from renewables. Until August 2010 local authorities were not allowed to sell electricity, except from CHP plants. But the coalition removed the restriction from all forms of renewable electricity. The coalition has also said that local government will be allowed to retain business property taxes from newly approved commercial buildings and facilities such as wind farms for the first six years of operation, rather than hand revenues over to central government. This will provide local government with an incentive to provide planning permission.

Germany is a much less centralized country than the UK is. The German federal government passes framework legislation, which the 16 regions then have to implement. This gives the federal government considerable power: the main driver of German renewable policy, the Feed-in Tariff, is set at federal level. However, the upper chamber of the German parliament is made up of delegations from regional governments, giving them great control over national legislation. Regions are the main tier for most land-use planning decisions. They can also impose specific anti-pollution regulations.

Within the regions, rural areas are divided into districts, and below them municipalities. In urban areas districts and municipalities are combined. Municipal councils have the power to 'regulate on their own responsibility all the affairs of the local community within the limits set by law'. Many municipalities use this power to justify expanding economic infrastructure and providing energy and gas. Municipal finances include 'nuisance' taxes (for example on noise), plus excise and expenditure taxes. But most municipal revenue comes from the regional and federal governments.

The Netherlands has particularly complex constitutional arrangements. As in Germany, there is a chamber of the national parliament made up of delegates sent by the regional governments. This has to approve bills before they become law, although it cannot amend them. The Dutch regions and municipalities have elected councils, but the mayors, who chair the councils, are appointed by the central government. The provinces have the right of oversight of the municipalities, but have little direct involvement in energy (unlike in water, which they control).

Denmark has five regions with rather limited powers chiefly to do with transport and hospitals, and 98 municipalities that provide or coordinate water, waste disposal and heating services. An exception to this general rule of decentralization is land-use planning in Sweden. This is decided by administrative boards, which are appointed by the central government.

In both Europe and the US, the issue of which tier of government should do what often dominates political debate, squeezing out the 'less important' issues of whether what is being proposed is sensible or necessary. In the UK, anything 'Brussels' wants to do is assessed primarily on the grounds of national interest and whether there is any justification for Europe to do this rather than the British government. Other countries take a more rational approach to EU politics, though the euro crisis and the need to bail out Greece, Ireland and Portugal (and quite possibly other countries before this book is published) has led to a noticeable decline in *communitaire* spirit in key countries such as Germany.

The EU does have the notion of 'subsidiarity' written into its treaties. This says that a higher tier of government should only take action if this would be more effective than lower tiers doing so. But this obviously leaves lots of room for political argument about what it means in practice. UK politicians like to identify 'red lines' before they enter into any European negotiations – issues on which no compromise will be considered. Anything related to taxation is always such a red line. However effective an EU tax would be, the argument goes, tax is a matter for elected national governments, not the 'Brussels bureaucrats'. This attitude led to the defeat of Commission President Jacques Delors's attempts to introduce a carbon/energy tax in the 1990s (see Box 4.1).

British politicians are prepared to let the EU do some things, provided that they are things that they approve of. The British prime minister who signed up to the 1986 Single European Act, giving up the British veto, was Margaret Thatcher, who was certainly no fan of Brussels. Her enthusiasm for increasing trade meant that she was willing to give up the national veto. UK foreign secretary William Hague, also no fan of Brussels, said in a speech in 2009 (while still in opposition) that the EU should focus more on important issues such as climate change, and less on endless new treaties. He read the speech out, so it didn't come from the heart, and he didn't respond to any questions he was asked about climate change. But at least he was prepared to let some pro-EU words pass his lips.

Box 4.1 Jacques Delors's carbon/energy tax proposal

In 1991 the European Commission proposed a carbon/energy tax, to be introduced at $3 per barrel of oil equivalent (around €6/tCO$_2$) in 1993 and rising $1 each year until it reached $10 per barrel of oil equivalent (around €20/tCO$_2$) in 2000.

The tax would have been levied equally on each fuel's carbon and energy contents. This was partly to encourage energy efficiency rather than fuel switching, but partly also to try to avoid the accusation that Delors, a Frenchman, was simply supporting nuclear power (which generates about 80 per cent of the electricity in France). A carbon tax would have worked to the advantage of French businesses. Nuclear power would not have paid the carbon tax, as it is low carbon. Renewables, also low carbon, would not have paid the carbon tax, but would also have been exempted from the energy tax, on the grounds that renewables were an 'infant industry' and so in need of special financial support.

The Commission proposed that national governments could keep all the revenue they collected. This was an attempt to get around the subsidiarity objection. There would be an EU-wide tax, but national governments would get the money. The Commission suggested that the money should be used to decrease other taxes, not increase spending, but this was merely a suggestion. National governments could have done whatever they wished with the money.

But the attempt to avoid national opposition failed. The UK, predictably, led the opposition, arguing that taxes were a matter for London, not Brussels. Other countries opposed the proposal on economic and industrial grounds, notably Spain, which was worried about the impact on its coal industry.

In 1994 Delors withdrew the proposal.

The fact that national governments select the commissioners adds a further complication to relations between Brussels and member states. The commissioner dealing with foreign policy, Cathy Ashton, is Labour and was selected by the UK's Labour government. New governments are not able to change commissioners until the commissioners' four-year term of office is up. Even when they come from the same party, the fact that they're sent to Brussels may be a form of being kicked upstairs. Günther Oettinger, the current energy commissioner, was previously head of the regional government of Baden Württemberg. He is from the same party as Merkel, the Christian Democratic Union (CDU). However, in the view of the American Embassy in Berlin – which we now know thanks to WikiLeaks[3] – his move to the Commission was not due to Merkel's admiration for him, but rather to 'remove an unloved lame duck from an important CDU bastion'. The cable goes on to say that Berlin has 'a time-honored tradition of sending unwanted politicians to the EU Commission'.

Despite the complexities of EU politics, the European Commission is extremely active on energy issues. This is in part to try to increase energy security, and in part to improve Europe's economic performance by using energy more efficiently. But it is also partly motivated by the need to reduce greenhouse gas emissions.

Box 4.2 EU energy policies

Soon after the Single European Act, the EU passed a directive to combat acid rain. This will lead to the closure or reduced usage of many coal power stations that emit too much sulphur or nitrogen dioxide in 2015 (which is progress, albeit almost three decades late).

Europe has subsequently adopted directives to establish minimum energy efficiency standards for many products and to label other appliances to show their energy efficiency. The Commission estimates that by 2020 the minimum performance standards it has set will deliver energy savings equivalent to 12 per cent of the electricity consumption of the EU in 2007. Standards for electric motors will deliver most, then televisions, then lighting (both domestic and street).[4]

In March 2007, in a genuinely surprising meeting, the heads of EU governments, reacting in Brussels for one of their frequent summits, used the directive on minimum standards to phase out tungsten filament lights. For good measure they also agreed to increase the amount of renewables to 20 per cent by 2020 and to reduce greenhouse emissions by 20 per cent. Prashant was an official in the UK working on energy efficiency and renewables policy at the time and the sneaky decision took everyone by surprise – the catchy soundbite 20/20/20 rather than analysis seemed the main rationale.

The labelling directive has increased the proportion of energy-efficient appliances being bought: in 1998 only 10 per cent of washing machines bought were in the most efficient category. By 2005 this had increased to 90 per cent.

There is also a directive aiming to make buildings more energy efficient. This mainly applies to new buildings. But new buildings account for only a small percentage of the building stock. Most of the houses, shops and factories that will exist in 2050 already exist today, and most of them use energy wastefully. The directive was strengthened in 2010; it now requires that all buildings undergoing major renovation will have to meet minimum energy performance requirements. But these standards are to be set by member states, so may be very weak. Some countries have done well on building standards and renovation. Germany already requires that any building undergoing substantial renovation must meet high energy efficiency standards. Sweden has gone further: every time a building is sold or rented out it must meet high efficiency standards.

In 2004 the EU adopted a directive on CHP, and in 2006 a directive on energy services. These are very weak. The countries doing best on CHP – Denmark, Finland, Austria, the Czech Republic – were doing so well before 2004. Significant developments on energy services also occurred before that directive. In 1998 the regional government of Upper Austria became the first public authority to use the ESCO approach for energy efficiency (see Box 6.3).

As well as these directives, the EU set up the world's first international cap-and-trade scheme, the Emissions Trading Scheme (ETS) in 2005. But so far the ETS has had little impact on emissions. National governments were in charge of allocating permits, and most governments allocated too many permits, so carbon prices were too low to have a significant effect on either energy use or investment in low-carbon electricity generation. Also, permits were given out for free rather than being sold or auctioned. From 2013 the

Commission will have control of allocation and permits will be auctioned to some sectors, including electricity generators. These are sensible improvements in the operation of the ETS. However, industrial activity will not reach pre-recession levels for many years, so emissions will be lower than anticipated in the baseline scenarios used to decide allocation levels. So the total number of permits allocated is still likely to be over-generous, driving the price of permits down.

The EU is very keen on setting targets on climate and energy policy (and indeed lots of other things). Its climate policy is now framed by the 20/20/20 package: emissions must be reduced by 20 per cent, 20 per cent of all energy must come from renewables and there must be a 20 per cent improvement in energy efficiency, all by 2020.

If there were no worries about carbon emissions, the easiest and cheapest way to increase Europe's energy security would be to burn lots more coal.

The Politics of Energy Policy

So how do politicians attempt to use the powers available to them on energy policy? Are there major differences between parties and political traditions? In the UK, there haven't been historically. The 1956 Clean Air Act, which banned the burning of coal in urban areas, was introduced by the Conservative Party. Nowadays the Conservatives are portrayed as being hostile to regulation but at that time the notion of 'one nation Conservatism' was more supportive of regulation. Government intervention in markets to promote social goods was widespread. Indeed in the 19th century it was the Liberal Party that was more hostile to regulation.

In most other European countries there is also little left/right divide on energy policy. But the absence of a major ideological divide hasn't stopped the arguments. Most politicians see argument as the central part of their job description. The one area of energy policy where there is no real controversy is energy efficiency. Almost all politicians agree that it makes sense not to use unnecessary energy. The problem is that this rather obvious point doesn't make a good political narrative. Jimmy Carter advised people to wear an extra jumper in winter – sensible advice but not sensible politics.

Issues about where the energy should come from are seen as much more interesting. Carter put solar panels on the White House roof. Reagan had them removed. Carter was president before climate change was a major political issue; his actions were intended to increase energy security. Since then, climate has emerged as the major issue dominating political discussion of energy. In the US, there is a substantial division between Republicans and Democrats on climate issues. Clinton/Gore tried to address climate change, but Bush/Cheney did not. Both Bush and Cheney had worked in the oil industry and got much campaign money from oil companies. Nevertheless, under Bush wind farms in the US expanded very substantially. This was not presented by the administration as necessary for climate reasons; it was to increase energy security. Under Obama the wind expansion has continued, and the climate benefits are now highlighted, but the actual policies have not changed substantially.

Not all Republicans are opposed to action on climate. When he was governor of California, Arnold Schwarzenegger said that he was strongly committed to climate action. John McCain tried to get a cap-and-trade bill through Congress before he ran for president. But selecting Sarah Palin as his running mate rather undermined his climate credentials (and indeed his credentials for being a good judge of character). Palin and the Tea Party do not accept that there is any human role in climate change. They argue that global warming is a socialist plot to overthrow capitalism and establish a world government.

Obama's White House energy adviser is Carol Browner. She is very experienced, having been head of the Environmental Protection Agency for the whole of Clinton's term of office. But she is also very party political, having called the George W. Bush administration 'the worst environmental administration ever'.[5] This lays her wide open to attack from the right for being – to use one of the strongest American insults – a socialist. A website called Global Warming Hoax – which accurately sums up its balanced approach – published an article soon after Obama picked her called 'Obama climate czar has socialist ties'.[6] Shock horror.

Climate change has not been such a left/right issue in Europe. Margaret Thatcher gave a strong speech on the need to control climate change in 1988. It is often thought, or claimed, that she did this in response to the UK Green Party getting 15 per cent of the vote in the 1988 European Parliamentary elections. But the speech was before the elections. Tony Blair talked a lot about climate change, though delivery of good policies was much less apparent. David Cameron used the issue as a way to 'decontaminate the brand' of Conservatism, but is also very knowledgeable and committed. The Conservative, Labour and Liberal Democrat parties had broadly similar manifestos on energy policies for the 2010 General Election. There are many 'climate sceptics' in the Conservative Party, but none of the leaders take this view.

In most other European countries, the existence of human-induced climate change is also not a cause of division between major parties. An exception is The Netherlands, where the far-right Freedom Party dismisses climate change as left wing and unproven. The Dutch government is a minority government that has an agreement with the Freedom Party to allow it to govern. But the two centre-right parties actually in the government are committed to climate action. Their coalition agreement states that Dutch energy policy will be guided by the European 20/20/20 agreement.

Renewables

Even those who don't accept global warming recognize that it makes sense to harness renewable energy for energy security reasons. As noted above, Bush did this. In Germany there has been a large degree of cross-party consensus on renewables. A Social Democratic Party (SPD) government established the Feed-in Tariff, and this continued largely unchanged (though rates were varied) through the SPD/Green and grand coalition governments. It still exists under the CDU/Free Democratic Party (FDP) government. But the FDP is economically liberal as well as socially liberal and so ideologically hostile to government subsidies. So the rates for support for renewables, particularly solar photovoltaic, are being reduced.

In the UK, there is little conflict between political parties about the need to harness renewable energy, nor even about the mechanisms needed to do so. A Conservative government introduced the Non-Fossil Fuel Obligation in 1990. Labour changed this to the Renewables Obligation in 2000. This requires electricity suppliers to source a rising proportion from renewables, so is similar to a US states' Renewables Portfolio Standard. Labour added a feed-in tariff for small facilities (anything up to 5MW) in 2010. The Conservative/Liberal Democrat coalition says that it will keep both the Renewables Obligation and the feed-in tariff.

However, there is substantial disagreement about where wind farms should go. And to the frustration of national politicians it is local politicians that decide whether to give planning permission to most wind turbines. Local councils are more vulnerable to 'not-in-my-backyard' (NIMBY). The Conservatives are stronger in rural areas, Labour in towns and cities. The Liberal Democrats span urban and rural. So sites where developers propose wind farms tend to be in Conservative areas, in England at least. (The Conservatives are weak in Scotland and relatively weak – though strengthening – in Wales.) So Conservative councillors and members of parliament are much more frequently lobbied by NIMBYs. Many Conservatives therefore oppose onshore wind farms. This is not the Party's formal policy, but is the view of some leading figures. Kenneth Clarke, former chancellor and now justice secretary, said in an unguarded moment before the last election that all wind farms should be offshore. Ed Miliband, the then energy and climate secretary, picked up on this immediately and demanded a retraction. Miliband's key point was that the UK has an EU obligation to get 15 per cent of its energy from renewables by 2020, and that reaching this will require lots of onshore turbines. Clarke is a pro-EU Conservative (a relatively rare beast), so there was clearly scope for political embarrassment. But energy wasn't part of Clarke's portfolio (he was shadowing justice too), so the retraction was made by the then shadow energy and climate secretary, Greg Clark. The retraction was made without delay, which is encouraging as it suggests that the Conservatives thought that it would be bad politics to be seen as anti-wind energy, despite the NIMBYs. Presumably they had read the polling evidence showing that most people support wind farms, but make much less noise in support than opponents do against.

David Cameron has moved from condemning wind turbines as 'bird shredders' to supporting them and putting a small one on his house. Greg Clark was very strong and effective as shadow energy and climate secretary (and in government is a minister in the Department for Communities and Local Government in charge of localism, so we hope he finds time to read our book).

Fossil fuels

Most politicians say that they support renewables. But policies and politics around fossil fuels are much more controversial. Obama has tried to introduce a cap-and-trade approach to greenhouse gases, but has been unable to get this passed by the Senate, where it was blocked by Republicans even before the mid-term election.

Previous UK Conservative governments, from 1979 to 1997, hastened the closure of coal power stations, but not for climate reasons. It was to reduce the power of the coal miners' union. The subsequent 'dash for gas' led to many large-scale gas power stations, with almost all the heat wasted up cooling towers. The current coalition has promised an emissions performance standard to limit emissions from any new coal stations. Labour says that it will support this, though while in office it rejected Conservative attempts to get one introduced.

In Germany, the CDU, FDP and SPD all recognize the need to cut emissions from coal stations, and support the use of CCS technology. The Greens reject CCS as insufficiently radical, arguing that all effort and money should go into renewables. The Dutch government is prepared to consider CCS, but only if there are strict safety standards and if local public support can be achieved. Local public support for a proposed CCS plant in Rotterdam, with the CO_2 pumped into an old gas field under one of the suburbs, proved impossible to obtain, so the project was withdrawn. Future Dutch proposals will therefore probably involve transporting the CO_2 to offshore gas fields, of which Holland has lots. The Dutch government's coalition agreement states that CCS projects will only be allowed after a licence has been granted for a new nuclear power station. This was probably a reaction to the strong public opposition to the Rotterdam CCS proposal.

Nuclear power

The energy issue that tends to divide European political parties from each other is nuclear power. Germany is a strong example of this. Until 2010 the right-wing, centre, left-wing and Green parties all agreed that Germany's nuclear power stations should be closed down by 2022, well before the end of their design life. But Merkel's government announced in September 2010 that it would allow nuclear power stations to remain open to the end of their design life. The SPD opposed this, as did the Greens. Anti-nuclear feeling remains very strong in Germany – around 100,000 people marched to the German parliament to protest against this new policy. The SPD argued that the constitution should be amended to require a referendum on this issue. They and the Greens also planned to challenge the government decision in the constitutional court, on the grounds that the Upper House of the German parliament, which is made up of delegates from regional governments, would not have a vote on the change of policy.

Merkel did not say that new nuclear power stations would be allowed. Her line was that nuclear power, like CCS, was a necessary bridge technology until Germany can be 100 per cent reliant on renewables. And she abandoned her attempt to keep nuclear stations open after the accident at the Fukushima I Nuclear Power Plant in Japan in March 2011. There is much talk of a nuclear renaissance in Europe, but new plants are only being constructed so far in France and Finland. Both are costing much more than expected (as nuclear power always does). In the UK, Liberal Democrat leaders have said that their opposition to nuclear power is not 'theological' (by which they presumably mean based on environmental issues – interesting that Liberal Democrats regard being green as a type of religion) but based on economics. The coalition agreement between them and the Conservatives states that new nuclear power stations should be permitted but without public subsidy. They have not formally defined subsidy. It does not include a floor price to the ETS, which the coalition has said that it will introduce, even though this will give financial benefit to nuclear, as to other low-carbon options. The UK Labour Party is also pro-nuclear.

The coalition agreement for the new Dutch government also states than an expansion of nuclear power is necessary. Three European countries have held referendums on nuclear power, all of which voted against. Austria had a referendum in 1978, while a nuclear station was under construction. The result was 50.5 per cent against and 49.5 per cent in favour. So construction was stopped. Sweden voted to shut down its nuclear plants, though none have actually been closed early. Italy did close its nuclear plants after a referendum. Berlusconi has decided that new ones should be constructed, but this is on hold after the events in Japan in 2011.

Despite Three Mile Island, the nuclear issue seems less politically controversial in the US than it is in Europe. Obama and Chu have given loans to companies wanting to build new nuclear plants. This has not caused significant political uproar.

Why Politicians Aren't Delivering More

Some politicians are stupid or wicked or corrupt. Some are all these things. But most aren't. However, there are several reasons why they don't deliver decently on energy issues. Most politicians are too focused on the short term – the time before they next have to face the electorate – and getting media attention. There is too much attention given to targets and timetables. Coherent policy is often blocked by rivalry between departments of government. Too much attention is paid to keeping constituencies within the party happy, rather than the general interest. The fossil fuel industry is very rich and very powerful. The transformation to a low-carbon energy-efficient economy will not be cheap. Many people in society resist most forms of change. And the dominant political tradition in Europe and North America stresses individualism: telling people (or even companies) what to do is widely seen as inconsistent with this ideology. The rest of this chapter considers each of these issues.

Short Termism

The electoral terms of politicians often stifle innovation. Facing an election in six, four or even two years, they are tempted to weigh current costs against only the savings over their term in office – and few investments pay off that rapidly.

The fact that the European Commission is not elected can have its advantages. It is pressing for an early end to the subsidies paid to member states' coal industries. The German government is resisting because lots of people work in the coal industry in Germany. Spain is the worst culprit and is actually trying to increase it subsidies to coal (and prioritize Spanish coal, despite the European single market). The members of the European Parliament are trying to delay the phasing out of coal subsidies. The fact that the two elected institutions, Council and Parliament, want to keep coal subsidies while the unelected Commission wants to end them is almost enough to make anyone doubt that democracy is consistent with sensible energy policy.

Soundbite journalism

All politicians love being quoted and photographed in the media. To be fair, it is an important part of democratic politics to get the message out to voters. But media obsession leads too many politicians to prioritize issues that they and journalists see as 'politically sexy' and downplay those regarded as less interesting or that don't provide good pictures. Hence the relative lack of emphasis on energy efficiency. Even in countries with lots of fossil fuels, it makes sense to use energy efficiently. There are lots of policies, and even more political speeches, about energy efficiency. But the speeches often don't get any media coverage. Not sexy enough. Ministers don't go to old buildings to cut the ribbon to open a newly double-glazed window. Opening a wind farm makes a much better photo.

Solar panels are often seen as even more sexy than wind farms. The UK's Deputy Prime Minister Nick Clegg opened a solar panel 'farm' in Sheffield (where his own constituency is) in the summer of 2010 and said 'It is a wonderful circular irony that the epicentre of the old industrial revolution is now serving as the epicentre of the new energy revolution'.[7]

Rearranging the deckchairs

It's often tempting for an incoming leader to look at the structures of government and find them wanting. 'Why is there a separate ministry of environment and energy when the two should so clearly be linked?', they ask, or 'Transport and urban planning are self-evidently best managed together'. This is made worse by the tendency of some heads of government to reshuffle ministers endlessly. In the UK, there was a different energy minister every year during the 13 years of Labour rule from 1997 to 2010.

Labour also had an unfortunate tendency to rearrange departmental structures. Climate change was first the responsibility of the Department for Environment, Transport and the Regions, then of the Department of Environment, Food and Rural Affairs, then of the Department of Energy and Climate Change. Departmentalism can be a problem: departments competing to be the top dog by holding onto budget, responsibility for policy, control over agencies. But constant changes in departmental structures made the institutions themselves, not just the individuals who ran them at a point in time, oriented towards quick solutions and responses to immediate crises, not to longer-term planning and investment. David Cameron sensibly ruled out any departmental reorganization when he became UK prime minister.

Similar departmental territoriality issues can arise at the local level as well, of course, but smaller offices simply have fewer layers of authority to convince to co-operate, so community-based initiatives may face a lower set of jurisdictional barriers than efforts launched under national auspices. However, many local politicians engage in wholesale reorganizing as a matter of imposition of the stamp of a newly elected government. For example, incoming mayors in New York City routinely abolish all departments as their first act in office. So short-termism affects local as well as national politicians. The State of Kentucky shocked political observers in the early 1990s by creating a 'Long Term Policy Research Institute'. The innovation was never picked up by other states – and fell victim to short-termism budget cutting in 2010.

Too much focus on targets

Much of the debate about how to tackle climate change has focused on targets and timetables. Targets and timetables have a role – they focus political and business and media attention. But they are much less important than actual policies to deliver emissions reduction. In 1994 Stephen was secretary of the Labour Party's Policy Commission on the Environment, which produced a report proposing a target of a 20 per cent reduction in UK CO_2 emissions (from 1990 levels) by 2010.[8] This report was published just before Tony Blair became Labour leader, so he was unable to change it even if he had wanted to. He did subsequently ignore many of the policy proposals, but never wanted to abandon the target. Indeed it was included in the 1997 Labour manifesto, so then became government policy.

One of the attractions of targets for politicians is that they can be set for dates when the politicians will not be in office – often referred to as 'not-in-my-term-of-office' or NIMTO. It is fair to say that in 1994 Labour did not expect to be in office from 1997 to 2010. Had Labour won the 2010 election, it would not have introduced emergency measures to meet the target. It would simply have set another target, probably for 2020. This is what they did on peat. In 1999, Labour had said that by 2010 peat would be eliminated from 90 per cent of the compost sold in the UK. 2010 arrived and almost half of UK compost still contained peat. So Labour set another target, this time for a complete phase-out of peat from amateur gardening by 2020, and enlisted a TV star to try to persuade the public to stop buying peat. This was an utterly feeble response – classic NIMTO. No use of government powers to achieve progress, just some more exhortation.

Labour did get some things right though. The 2008 Climate Change Act set national 'carbon budgets' – the amount of CO_2 to be emitted over defined periods – that are likely to be more effective than long-term targets. Of course, the Act also has the headline target: a 60 per cent reduction – subsequently increased to 80 per cent – in greenhouse gas emissions by 2050. 2050 is more likely to be 'not-in-my-lifetime' for most politicians, rather than NIMTO. But it also set carbon budgets to be met much earlier, some for this decade, and created a Climate Change Committee to advise on how the budgets could be met. The Committee is not supposed to comment on policy but inevitably it does. Its first chairman is Adair Turner, who is strongly committed to radical action on climate – an early Committee report called for a step change on decarbonization. He used to be head of the Confederation of British Industry, and so can't credibly be accused of being anti-business.

Targets and timetables also dominate international negotiations. The Kyoto Protocol set targets for greenhouse gases, to be reduced from 1990 levels by 2008–2012. It has achieved a little. Emissions from countries that ratified them were lower than 1990 levels – though not by nearly enough to control climate change. The US, which signed but never ratified Kyoto, has allowed its emissions to rise by almost 15 per cent.

However, it is hard to be too positive about the impact of Kyoto. The host country Japan accepted a target of a 7 per cent reduction but by 2006 its emissions had risen by over 5 per cent. Despite being signed in 1997, the Protocol only came into force in 2004 when 55 per cent of the developed countries had ratified it, and the country that took it past this finishing post was Russia. But Russia, with an abundance of cheap carbon allowances following the collapse of its economic output following the collapse of communism, is destabilizing the carbon market by flooding it with 'hot air'. The EU will not meet its Kyoto target without the purchase of this hot air.

Kyoto was a legally binding agreement – a phrase that has become something of a mantra among campaigners. A binding target has more effect than an aspirational one, but it does not guarantee achievement. Several EU countries will miss their share of the EU's Kyoto target. (In EU jargon, the sharing out of targets is usually referred to as 'burden sharing', which is a very unfortunate use of language. Investing in secure future energy sources shouldn't be seen as a burden.) In 2008 Spain's total greenhouse gas emissions were 40 per cent above the 1990 level; its target is to be only 15 per cent above. Austria, Ireland, Italy and Denmark also have virtually no chance of meeting their Kyoto targets. The Commission will be able to fine them, but the prospect of fines has not spurred the governments to take the targets seriously.

Constituencies inside the party

Political parties aren't single blocs. They're made up of different interests and different individuals. Party leaders have to keep internal constituencies reasonably happy, as well as appealing to the wider public.

A clear example of the power of party constituencies is the fact that the UK's Climate Change Levy is not, despite its name, a carbon tax but an energy tax. After Labour won the 1997 General Election Stephen spent two years as adviser to Environment Minister Michael Meacher, in the newly created Department of Environment, Transport and the Regions. Deputy Prime Minister John Prescott was the head of the Department. Prescott and Meacher were quite an effective team: Meacher excellent on policy detail, Prescott good at getting political agreement.

Labour's manifesto had promised to increase revenue from green taxes and reduce taxes on employment, and proposed to introduce a tax on commercial and industrial use of energy. Stephen had published a book calling for this in 1996,[9] so tried to be as involved as possible in the discussions about what the government should do (though departmentalism is particularly strong on issues of tax – no one is supposed to talk or even think about tax except the finance ministry, Her Majesty's Treasury).

It was clear that the government would introduce such a tax. There was no significant internal party opposition and Labour had a large majority so wouldn't have any problem getting it through Parliament. But the main question was whether it should be a carbon tax or an energy tax. Prime Minister Blair was already taking a serious interest in climate change and would have preferred a carbon tax. The climate case for taxing carbon rather than energy was pretty clear. But the other – unspoken – advantage in the view of Blair and his staff was that this would be good for gas and nuclear electricity generation and bad for the coal industry. This had political advantages in Blair's eyes because his wasn't a Labour government, it was a 'New' Labour government, not in hock to old industries, traditions or trade unions.

The secretary of state for trade and industry, in charge of energy, was briefly Peter Mandelson (until his first resignation in 1999). Mandelson was quintessential New Labour and no fan of the coal industry, so he supported a carbon tax. John Prescott has never defined himself as New Labour. He is simply Labour, close to trade unions and proud of Labour's traditions. Blair could never have managed the Labour Party without Prescott's help. Prescott opposed a carbon tax as this would be bad for the coal industry.

For most of Blair's premiership, the government had in effect a 'big three': Blair, Prescott and finance minister Gordon Brown. On issues on which Blair and Brown disagreed (i.e. most issues), Prescott was the de facto referee. But on the issue of a carbon tax, Brown had the casting vote. Brown was never particularly engaged with climate issues, or anything that he regarded as 'green'. He did regard himself as New Labour, but was never averse to playing to particular constituencies within the Party. He did, after all, want to become Labour leader and prime minister. So Brown came down in favour of an energy tax rather than a carbon tax. Thus the die was cast.

It was not, however, necessary for any politician to admit that an energy tax had been chosen to protect the coal industry. Many environmentalists also favoured an energy tax rather than a carbon tax. Stephen has to accept some responsibility on this, not that he had much influence over tax issues while in government. His 1996 book had argued for an energy tax rather than a carbon tax to avoid simply promoting fuel switching. Switching from coal to nuclear would mean simply swapping one set of problems for another, reflecting his then opposition to nuclear power. What was really needed, he'd argued, was a greater focus on energy conservation and renewables.

The power of the fossil fuel industry

Most of the energy used in the world today comes from coal, gas and oil. This makes the fossil fuel industry very rich. Some of its revenue is used for political donations to candidates thought to be in favour of continuing fossil fuels' dominance. These donations are not exactly bribes – few politicians are corrupt. But a politician who speaks and votes against the interests of fossil fuel companies cannot expect future donations from them, and clean energy companies are not yet wealthy enough to step in and replace the cash.

Political funding is a particular problem in the US, which has no public funding of political parties and where contesting elections is very expensive. In the 2010 gubernatorial and Congress election, business is reported to have spent $4 billion nationally on promoting their candidates.[10] Only the ultra-rich (like Meg Whitman, ex-CEO of eBay, who spent $140 million of her personal money in the California gubernatorial election) can completely side-step corporate money.

Funding by special interests restricts what is politically possible. Clinton/Gore signed the Kyoto Protocol in 1997, but never sent it to the Senate for ratification, because they knew it had no chance of passing. Before the Kyoto convention, the Senate had passed a statement by 95 votes to nil saying that it would not ratify a protocol on greenhouse gases that did not include targets for major developing countries such as China and India. Since Kyoto did not include such targets, the treaty had no chance of being ratified. George Bush formally abandoned the Protocol; he didn't believe the Protocol was good for the US and this position probably did him no harm in securing enormous campaign contributions from fossil fuel companies such as ExxonMobil. Had Al Gore become President in 2000, he would not have been able to get the treaty ratified either. Too many senators disagreed with the Protocol, including some who are concerned about the threat of climate change.

The dominance of fossil fuels also gives these industries enormous indirect influence. Politicians are not only concerned about jobs for internal party reasons. Most politicians are concerned about keeping unemployment as low as possible, and those who represent coal areas also have concerns about the jobs of their electorate and the economy of their constituencies. In aggregate, the Organisation for Economic Co-operation and Development (OECD) countries pay $400 billion a year in subsidies to fossil fuels, according to the IEA.[11] Around 175,000 people work in the US coal industry, in mining, transportation and coal power plants. West Virginia and Kentucky have the highest number of workers in coal.[12] The coal industry also still provides tens of thousands jobs across the EU. EU countries (particularly Germany and Spain) paid out €3 billion in national coal subsidies in 2008. Reductions in subsidies to coal or taxes and regulatory controls on coal's use would reduce employment in the coal sector, making such policies very unpopular with many politicians.

Energy efficiency and renewable energy are also capable of creating employment. In some countries it already does – Germany has over 150,000 working in renewables. But this number is not high enough to make the German government want to end subsidies to coal.

It isn't only the fossil fuel industry that lobbies against sensible energy policies. The most effective way to create jobs on climate and energy issues would be a large programme of energy efficiency work, improving the existing building stock. This would create work for builders, particularly hard hit by the recession and fiscal deficits. It would also reduce fuel poverty, saving thousands of lives each year. But the construction industry isn't keen on being told what to do. The European Commission wanted to mandate that any property being retrofitted be made energy efficient, but extensive lobbying meant that this only appeared in a very watered down version in the 2010 buildings directive.

The costs of transformation

Politicians are better at giving out money, when they have any, to support new industries than they are at reducing subsidies to old industries. OECD countries give $27 billion a year to renewables, according to the IEA.[13] This sounds a lot, but is much less than the amount needed and is only 7 per cent of the $400 billion every year in subsidies to fossil fuels.

Clean energy is at present more expensive than fossil fuel and so needs some form of financial support – taxes or cap-and-trade schemes on dirty energy, obligations on energy companies to get a certain proportion of energy from clean sources or direct subsidy. Energy efficiency programmes also cost money.

The cost of the low-carbon transformation is a major barrier to progress. Most governments have enormous budget deficits. Green taxes are one way to increase government revenue. So are cap-and-trade schemes where permits are auctioned. These approaches are often condemned by opponents as 'stealth taxes', though all have to go through congress, parliament or local councils, so they are not that stealthy. But new or increased taxes are never popular, and they are particularly unpopular at present in the US – the Tea Party has a simple bumper sticker slogan: 'people are Taxed Enough Already'.

Another approach is to raise money through levies on consumers, or place obligations on energy companies, which then feed through into higher energy tariffs. This has some political appeal: there is not yet a political movement called Levied Enough Already. But policy instruments that raise money or impose obligations are de facto taxes, whatever they're called. In the UK the Office for National Statistics has now officially defined the Renewables Obligation as a tax. Californian voters have passed a 'proposition' that will require a two-thirds majority in California's Congress for any measure that would result in any taxpayer paying a higher tax. This is being used by the California Chamber of Commerce to argue that the cap-and-trade – in their eyes a tax – needs a two-thirds majority.

Those opposing any change

Another blockage to political progress comes from those who oppose any change in the look and feel of a community. This has been the main barrier to renewables in the UK. Planning laws have prevented the urban sprawl and ribbon development that blights so much of the US. But it also gives significant opportunities to NIMBYs to stop most proposed onshore wind farms developments.

The fact that three-quarters of planning applications for onshore wind farms going to local councils are rejected could be seen as a triumph for local preferences. But it isn't really. Poll after poll shows that the majority of people living in the vicinity of new wind plants support them. Between 70 and 80 per cent of the UK public support wind farms, only around 10–15 per cent actually oppose them. The figures barely flutter when respondents are asked about a wind farm being erected near their home. Support is highest in communities that actually have a wind farm nearby. Familiarity breeds content, not contempt. The following comment is not atypical:

> I would like to place on record how opposed to the wind turbines I was. I even had a letter in the Yorkshire Post against them and that was before they became a reality. However they have turned out to be if anything an 'asset' to the area, both aesthetically and environmentally. We could really do with having a lot more of them, as they are adding to our energy in a green way, and look very majestic.
>
> A defector who used to oppose wind farms[14]

Measurements confirm that the sound of a wind turbine 350 metres away is less than the normal background sound in a busy office,[15] impact on wildlife and birds negligible compared to the death toll wrought by cars and cats. But those in favour of a wind farm, or indeed anything else, tend not to speak out. Opposition is sadly a greater motivator than is support. In the UK some green NGOs have run a campaign called 'Yes2Wind' to try and whip up some enthusiasm, but without great impact. No fuel source is invisible, and open-cast coal mines aren't exactly pretty. But they're out of sight of the NIMBYs.

Some conservationists also oppose energy efficiency measures. Parts of the UK are designated 'conservation areas' and have area committees to campaign against any 'unsightly' developments. Conservation area committees dislike any change to buildings – certainly wind turbines on roofs and solar panels visible from the street, external insulation and double glazing.[16]

Oxford University's Environment Change Unit estimates that around 5 per cent of UK homes are in conservation areas. Around 370,000 are listed by National Heritage, which further restricts changes that may be made to the inside or outside of the building.[17] In the US, programmes to retrofit old homes using stimulus money are being held up while the conservation lobby considers each application.

There are some grounds for optimism regarding large conservation organizations. Key US conservation groups the Audubon Society, League of Conservation Voters, Natural Resources Defense Council, National Wildlife Federation and Sierra Club campaign actively in favour of energy efficiency, renewables and CCS. In the UK, the Royal Society for the Protection of Birds, which has over 1 million members, now says that most wind farms should be supported, as the greatest threat to birds is climate change. Natural England, the public body responsible for preserving England's landscape, also says that most wind farms should be supported.

A final reason why politicians are not doing better on energy issues relates to the dominant political ideology in both North America and Europe. Liberal democracy, with its close ties to capitalism, is built around a high degree of libertarian individualism, a fact more starkly evident in the US than elsewhere but still relevant in other countries.

Regulatory policy is specifically intended to curtail individual and business 'liberty' for the greater good. The prevention of harm to others is the central principle of law and order policies, and entirely consistent with liberal democratic theory. So opposition on grounds of defence of libertarianism is often a cover for opposition based on financial interest or other forms of self-interest. Politicians seeking progressive change need greater courage of their convictions. But too often lack of political bravery prevents much progress.

Can Democratic Politicians Ever Deliver?

Even the best democracies agree that when a major war approaches, democracy must be put on hold for the time being. I have a feeling that climate change may be an issue as severe as a war. It may be necessary to put democracy on hold for a while.

James Lovelock (2010)[18]

The country that has the greatest installed wind power capacity is now China. The Chinese government doesn't need to worry about NIMBYs. It just needs a map to decide where to put the wind farms. The Chinese government has said that China will be the world's largest manufacturer of electric vehicles by 2012, so it probably will. This will strengthen the Chinese economy in the decades to come. The Chinese government doesn't need to worry about inconveniences like elections, so isn't prone to short-termism.

So is Lovelock right that we can't afford democracy right now? Should we all be prepared to restrict or bypass democracy in order to make the necessary energy transformation. No. Democracy is, as Churchill so memorably and accurately put it, the worst form of government, apart from all the others tried so far. Lovelock is correct that climate change represent an enormous threat. This isn't a major issue of controversy in Germany, Austria, Scandinavia or The Netherlands, but is widely disputed in the US, and increasingly so in the UK. Politicians wanting to control climate change are therefore wise to play down the climate issue and emphasize other benefits: energy security, economic efficiency, health and comfort. All of these will improve with more energy efficiency and more use of renewable energy. Most people will be healthier and richer once the world has an efficient low-carbon energy system, not only than they would be if climate change is not controlled (which wouldn't be hard), but than they are today.

However, there is a substantial transitional cost. Renewable energy can provide all the world's energy, but it will take several decades to get to that very desirable goal. So other low-carbon bridge technologies (to use Angela Merkel's excellent name for them) are needed, like CCS and gas-fired CHP. Support for these technologies does have to be presented in climate terms. No other argument works.

Heads of government should stop the endless reorganizations of departments and reshuffling of ministers. The electoral cycle makes some short-termism inevitable, but there's no need to make it worse by constant change.

Politicians should play down targets and timetables, and focus instead on policy and delivery. Politicians must be prepared to cede powers to another level to help the energy transformation. Sometimes this will be to a higher level – for example, allowing the EU to set a minimum energy tax. But often it should be to a lower level. Local government must be repowered. However, local government can't just be given all the necessary powers and then left to get on with it. They must be properly resourced for delivery. And local government needs to be large enough to deal with necessary energy and economic issues. Many US counties and municipalities are too small. The need for local government organizations of the right size is the exception to the rule that there should be no more institutional tinkering.

Climate change and energy issues should not be seen as left/right political issues. Energy policy shouldn't be seen as a left/right issue. The market is a powerful and valuable tool for innovation and advance. But it can't be a free market. Competition is better than monopolies are at encouraging innovation, but it is not better at delivering a coordinated solution. If communities are to reduce their energy use and replace fossil energy use with sustainable energy sources, regulation is essential.

In designing regulations, governments should work with businesses. But politicians should be sceptical of industry-led technical committees. Well-resourced opponents of change are often effective at filibustering processes. Politicians need to decide and act for their citizens, sometimes against the interests of incumbent firms.

Cross-party consensus should certainly be encouraged. But it will never be easy, certainly when it involves taxation (which we think energy policy should). No energy policy has the unanimous support of the major constituencies within the country, so the policy is never 'settled'. Debate and discussion are healthy. Our political leaders should be happy to talk, but not for ever. After a decent time of discussion, they need to decide and show some leadership. That's what we elect and pay them to do.

Notes

1   Hobsbawm, E. (1995) *Ages of Extremes*, Abacus.
2   Buchan, D. (2009) *Energy and Climate Change: Europe at the Crossroads*, Oxford University Press.
3   Cable from American Embassy in Berlin, 31 December 2009, WikiLeaks, http://213.251.145.96.
4   European Commission (no date) 'Sustainable and responsible business', http://ec.europa.eu/enterprise/policies/sustainable-business/ecodesign/product-groups/index_en.htm.
5   New York Times (2011) 'Times topics', 24 January, http://topics.nytimes.com/top/reference/timestopics/people/b/carol_m_browner/index.html.
6   Global Warming Hoax (2009) 'Obama climate czar has socialist ties', http://globalwarminghoax.wordpress.com/2009/01/12/obama-climate-czar-has-socialist-ties.
7   *The Star* (2010) 'Clegg launches energy project', *The Star*, 23 August 2010.
8   Labour Party (1994) 'In trust for tomorrow', Labour Party.
9   Tindale, S. and Holtham, G. (1996) *Green Tax Reform: Pollution Payments and Labour Tax Cuts*, Institute for Public Policy Research.
10 Carter, T. (2010) 'Right-wing Democrat wins in California gubernatorial election', World Socialist Web Site, www.wsws.org/articles/2010/nov2010/cali-n05.shtml.

11 Sud Ouest (2009) 'International Energy Agency welcomes phasing out subsidies on fossil fuels', Sud Ouest.
12 Sourcewatch (2008) 'Coal and jobs in the US', www.sourcewatch.org/index.php?title=Coal_and_jobs_in_the_United_States.
13 Sud Ouest (2009) 'International Energy Agency welcomes phasing out subsidies on fossil fuels', Sud Ouest.
14 E.ON (2011) 'Public attitudes', www.eon-uk.com/generation/publicattitudes.aspx
15 Renewable UK (no date) Noise from Wind Tubines: The Facts, www.bwea.com/ref/noise.html.
16 Vaze, P. (2009) The Economical Environmentalist, Earthscan, London.
17 Environmental Change Institute (2005) 'The 40 per cent house', Environmental Change Institute.
18 Lovelock, J. (2010) 'Humans are too stupid to prevent climate change', The Guardian, 29 March.

Chapter 5

Delivering Low-Carbon Communities

The big lesson that I learnt in that first term was that actually today's politics is a lot more to do with structural change, project management and delivery than it is to do with ideological fixations, left versus right, or the notion that you can, by edict from government, change things.

Tony Blair (2010)[1]

No institution can possibly survive if it needs geniuses or supermen to manage it. It must be organized in such a way as to be able to get along under a leadership composed of average human beings.

Peter Drucker[2]

Many strategy, NGO and policy-maker types make the upsetting discovery that their great new public policy idea, which has been honed into a riveting business case, still never leads to actual improvements. The idea fails for a whole bunch of prosaic reasons that seemed to them boring details. Small details like delivery. The smartest ideas founder because insufficient thought is given to whether what looks good from a cost–benefit analysis actually makes any business sense.

Can the current structure of energy companies make profits out of low carbon and energy efficiency? This structure was established when the main public policy ambition was to make power and gas widely and cheaply available. The syllabus has changed and the exam question being posed is how to reduce demand and squeeze more useable energy from fuel. Energy efficiency in particular is a different product to gas and electricity; the businesses we are writing about will have to market the *idea* of using less energy. There is an adage in marketing that the marketer should know and understand the customer so well the product or service sells itself. Unpacking this statement, businesses can only do so much to change people; they have to take them as they find them. Businesses have to develop their wares around customers; they have to take account of all the weaknesses and foibles discussed in Chapter 8. These include customers' disinclination to think alike on some matters, their fondness for conforming in other matters, their time myopia, their poverty of resources, their poverty of imagination, their love for their homes, their social involvement or alienation, their financial and social capital and their aspirations.

There are several possible business models. The mission could focus on profit maximization, customer service or environmental protection. Businesses could be investor-owned, publicly owned or co-operative. We don't believe there is any single correct model but some models are more likely to deliver success than others.

Co-operatives

Consumer-owned energy co-operatives are not as conflicted as investor-owned companies. They exist to provide energy (and increasingly energy *services*) as cheaply as possible for the benefit of their customers. Often they have wider social and environmental issues built into their missions. The UK co-operative tradition[3] includes principles such as concern for the community, democratic member control (often one person one vote), education, training and information. Though these organizations are not as conflicted through their mission, there might still be issues of lack of knowledge, appropriate skills within organizations and institutional inertia.

How have co-ops adapted to the low-carbon agenda? Co-ops are an established feature in Danish energy markets. In 2001 there were some 430 municipally owned or co-operatively owned DH companies.[4] These arose from the enactment of the 1979 Heat Supply Act. This stipulates that homes had to connect to local heat networks. The creation of such heating monopolies could result in problems through inefficient operations or bad customer service, so the privileges of being conferred this monopoly had to be balanced with a requirement for a significant degree of consumer control, not-for-profit charging practices and transparent pricing to ensure consumers could verify performance. These stipulations put off many private sector actors but are necessary for the public to have trust in the business. The co-ops operate as open book organizations, so any member (customer) is allowed to scrutinize costs and revenues to ensure surpluses have been either ploughed back into the business or returned to customers. The 'monopoly' right to supply warmth is a great virtue when seeking finance. Investors can see predictable and stable cash flow, and this greatly reduces the risk they bear when investing in expensive heat networks.

Co-operative and local ownership is also a key attribute of the Danish wind industry. In the 1980s and 1990s the majority of wind turbines were co-operatively owned. At that time ownership was widely spread so many people, local people, had a stake in making the project a success. In recent years most have been part co-op, part private sector. The Middelgrunden Wind Turbine Cooperative has 8500 members. It is situated 2km off the coast of Copenhagen and is visible from the city. Because it is sited close to its market, electrical losses are low – just 2.7 per cent compared to the national grid loss of 9 per cent. There is also the less tangible benefit of heightening awareness and interest in energy efficiency and low-carbon energy. Since the 1990s, larger wind farms often sited offshore have become more important and utilities have become the main investor. In 2009 around 15 per cent of turbines were still co-operatively owned.[5]

The UK does not have great experience of co-operative energy production. Even so, the co-op model has been successful in gaining popular support for wind plants. Baywind Energy Co-operative was set up in 1996 in Cumbria to invest in local wind farms. Baywind supports local renewable generation and energy efficiency. The first share offer in 1996 raised £1.2 million to buy two turbines at the Harlock Hill wind farm. Two years later a second share offer raised funds for another turbine. Revenue was around £500,000 in 2009 and shareholders obtained returns of 8 per cent gross (enhanced to 10 per cent through tax breaks). The co-operative has donated £3000 towards a local school's photovoltaic panel and also supports local domestic energy efficiency measures. The co-op also makes grants and loans to other energy co-ops (notionally competitors!), supports local domestic energy efficiency measures and is developing proposals for new sites itself. Voting rights are distributed equally among the members, regardless of the number of shares held. The original board of directors included seven members from towns in Cumbria. Baywind has a minimum shareholding of 300 and a maximum of 20,000. Hence a stake in the co-op is within easy reach of most people but no single individual or organization can have a controlling interest. The co-op currently has over 1300 shareholders throughout the UK and abroad. The board is elected by the whole membership at the annual general meeting and is supported by a full-time paid administrator, who is also a director.[6]

Co-operative and community ownership of energy systems are particularly well suited to small remote communities that are close knit, and that can take advantage of grants available for remote regions. A couple of examples of remote Scottish communities that have embraced community energy are shown in Box 5.1.

This admirable approach – moving beyond the desire to maximize the bottom line – is common throughout the co-op (and public) power movement. This is well demonstrated by the example of the South San Joaquin Irrigation District (SSJID) in California (see Box 5.2). Jeff Shields, the general manager of this organization, is a rare beast: one of the only managers of a public entity in the state who is running a huge trading account surplus. SSJID owns and runs substantial hydroelectric power resources and has accrued $88 million of surpluses since 2004. He wishes to return these savings to his customers by acquiring the rights to distribute power and energy efficiency directly to 40,000 customers in his district through acquisition of the local power network. His team of around 30 staff already operate their hydro plant responsively to the system operator,

Box 5.1 Community-owned renewables on Scottish islands

In 2004 the Gigha wind farm opened.[7] Three 'second-hand' turbines with a combined capacity of 675kW. These now generate about two-thirds of Gigha's electricity need, save 900 tonnes of CO_2 per year and make an annual profit of £150,000, which is reinvested in housing renovation. This was the first community-owned, grid-connected wind farm in the UK. The island had been owned by a wealthy landowner and had been depopulating for a century, making the provision of basic services such as education and health increasingly unviable. But the community ownership of the island and its energy infrastructure has led to some growth and now 120 people live on the island and many tourists visit every summer.

Until 2007 people living on the communally owned Isle of Eigg[8] had to rely on diesel generators and small hydro plant for electricity. Its 37 houses and five commercial properties then switched over to a local renewables system, which consisted of four 6kW wind turbines, a 10kW solar photovoltaic array and a 100kW run-of-river hydro scheme. Residents have also improved insulation and have to live within an energy cap of 5kW for households and 10kW for larger properties. If they exceed this, the system locks them out of the grid and they have to pay a fine of £25 to be switched back in. Given the northerly location and the island's windiness this is quite a sacrifice. Eigg previously had two 6kW hydro plants, which have been integrated into the new island grid. There is a diesel back-up generator for when there is not enough water, wind or sun. The project cost £1.65 million, mostly from grants, with £93,000 from residents. This was expensive, but cheaper than the £4 million required to connect the grid to the mainland.

Emissions of CO_2 for heating and power have been cut from an already respectable 9.6tCO$_2$ per household to 6.2tCO$_2$ per household. The island residents won a £1 million prize in 2010 for their efforts at curbing emissions. They'll be using this to further reduce their emissions by replacing some of their coal and kerosene heaters with biomass heaters and improving energy efficiency. This might reduce emissions by a further third.

to maximize income from power sales. The water and irrigation business means they already have extensive billing systems. Shields argues, 'I have no interest in selling my customers more power. I want to sell them less power and would agree to cut consumption by 4 per cent a year, much more than the 2.5 per cent required by the regulator from the investor-owned companies'. His plans include working with the housing authorities to replace the ancient trailers the district still uses to house its social tenants with new highly insulated energy-efficient models. Not the type of investment most energy companies would consider, but perhaps correct for this area given the social pressures and needs of the local community. Co-ops are not restricted to small energy developments. There are some examples of large developments too, such as Copenhagen's DH networks and the 40MW Middelgrunden Wind Turbine Cooperative described above. But because of its hyper democracy, the co-op model perhaps is less practical for large complex tasks such as city-wide energy efficiency programmes.

Box 5.2 South San Joaquin Irrigation District

SSJID is a publicly constituted irrigation district that supplies irrigation water to South San Joaqin County near Sacramento California and drinking water to some citizens too. It is independent of the city and owned by local citizens. It has provided irrigation to farmers since 1914. In 1954 it built the Tri-Dam project, year-round water to its farmers and also 120MW of electricity generating capacity. Until 2004, power was sold through a long-term contract to the large investor-owned utility PG&E and the revenue used to repay the money borrowed to build the dam. Since 2004, SSJID has sold electricity to the wholesale power market. As well as hydro, SSJID owns a 1.4MW solar photovoltaic plant. It is continually investing in its hydro capacity too, upgrading the turbines in its hydro plant to further expand output. It sells 30 per cent of the renewable electricity to its own customers and sells the rest to the California market.

SSJID now wants to buy the local distribution business from PG&E and take over responsibility for power supply to 35,000 customers. PG&E and SSJID have not been able to agree a price – the price requested by PG&E is ten times greater than that being offered by SSJID (and 20 times greater than that booked in the regulatory accounts). The regulator has fined PG&E $404,000 for its conduct during the negotiations.[9] SSJID can compulsorily purchase the wires from PG&E if they cannot agree a price.

SSJID's desire is to provide its customers with more reliable power (PG&E is not especially bad), reduce tariffs and increase energy efficiency. It is offering to provide a 15 per cent discount over the PG&E prices.

Municipal Energy Organizations

Municipal energy organizations are also not as conflicted as investor-owned companies. The US has a long tradition of municipal power companies. The contrast between the missions and values of public and investor-owned companies are instructive. The two types of entity are fundamentally in the same business – providing energy and energy services to people. But their cultures are different. There is a strong public ethos within public utilities. In 2009 the total compensation for Alan Fohrer, CEO of the investor-owned Southern California Edison, was $3.2 million, nearly 11 times that of David Freeman the head of the neighbouring and publicly owned utility LADWP, on $300,000.

In 2006 the prices paid by household customers of investor-owned utilities were 14 per cent higher than public utility customers. Rural co-ops were 3 per cent higher than public utilities.[10] Similar differentials were found for commercial customers too. Despite the investor-owned companies' reputation for efficiency, they cost more than public and some co-operative utilities. This is partly because, as the next chapter explains, the cost of borrowing money for investment is lower for public entities than for private entities. But it is also because public and co-op utilities are forbidden from many forms of spending on advertising and political lobbying, salaries of the board room staff are lower, and there are no dividends to pay to shareholders.

LADWP's mission statement reads:

We are a publicly-owned utility committed to providing clean, reliable water and power in a safe, environmentally responsible and cost-effective manner with excellent customer service to the communities we serve.[11]

LADWP is largely fulfilling this mandate, but its power reliability has deteriorated over the past ten years as it has had to reduce investment. The city authorities are asking it to transfer 7 per cent of its revenues to them each year to help fund Los Angeles's budget deficit. Its reliability figures are better than those of two of the large Californian investor-owned utilities but worst than the third. Boards, rather than the state regulator, decide public power companies' tariffs and energy efficiency programmes. Box 5.3 on Sacramento Municipal Utility District (SMUD) – one of the best-managed public utilities in the US – gives an overview of the different programmes to encourage energy efficiency and community energy.

Southern California Edison's strap-line is 'leading the way in electricity'. It amplifies in its operating priorities: 'We operate safely, We meet customer needs ... We protect the Environment ... *We grow the value of the business*' (authors' emphasis).[12] The choice of verbs is significant: satisfying customers and the environment but growing the value of the business. A former climate minister in the UK used to draw a distinction between the departmental priorities that gave the ministry its 'licence to operate', and those such as the 'war on climate change' that were the ministry's legacy to the future. Compliance with environmental and health and safety measures provide investor-owned utilities with a mandate to operate – but they are not what motivates the CEO to get out of bed each morning.

How does a company with a fixed geographic service area like the typical US electricity utility grow in value? The answer is either by cutting costs or selling more energy; the latter is of course the opposite of what is desired from a sustainable energy perspective. The mission of the company is fundamentally at odds with a sustainable outcome. It is particularly pernicious since the utility has to make long-lived investments in wires and generation capacity, which is built on the anticipation of a certain level of demand. If the demand for electricity is

Box 5.3 Sacramento Municipal Utility District

Sacramento voted for SMUD to be established in 1923, but its first delivery of power was in 1946 when the California Supreme Court finally dismissed the incumbent PG&E's objections against the compulsory purchase of PG&E's Sacramento assets. The public utility now serves over 1 million people, making it the second largest public utility in California. SMUD operates as a community-owned agency instead of a department of a local government (as does LADWP). It is free to use its surpluses to reinvest as it sees fit. It is overseen by a board of seven directors elected by customers to represent wards within the city. Board meetings are televised and members of the public are allowed to address the board if they make a written request. SMUD's purpose is 'to provide solutions for meeting our customers' electrical energy needs' and its vision is 'to empower our customers with solutions and options that increase energy efficiency, protect the environment, reduce global warming, and lower the cost to serve our region'.[13]

SMUD's electricity tariffs are around 30 per cent lower than those in the surrounding areas and the percentage renewables in its generation is around double that of the California renewable portfolio standard. SMUD is pursuing a range of innovative initiatives to reduce demand and encourage photovoltaic panels (the only renewable electricity suited to the city).

The benefits of some of its programmes are hard to quantify and would therefore not be permitted by the energy regulator. One of its most popular demand reduction programmes has been 'Shade Trees'. SMUD provides small trees for free for residents to plant near their homes to reduce the need for air conditioning. The policy is also good for community morale and has resulted in the greening of spaces adjacent to buildings. It is widely used by other public power utilities.

It has also developed a careful customer-centric approach to rolling out smart meters and retrofitting homes to enhance public confidence in the technologies. SMUD maintains a list of qualified contractors: they check the contractors have the correct licences and offer appropriate warranties and have attended the Building Performance Institute training. These contractors act as effective spokespersons for the retrofit programme. SMUD also provides commercial loans for energy efficiency. So far 10,000 households have taken advantage, mainly for installing heating, ventilation and air conditioning (HVAC) systems and windows. All purchased items have to be at least Energy Star rated, and thermal performance of windows has to exceed Energy Star.

SMUD has been rolling out smart meters but much more cautiously than other companies to avoid the customer backlash that has accompanied some of the more aggressive programmes. SMUD encourages customers to voluntarily use a time-of-use tariff. Time-of-use tariffs are controversial. At peak time – for instance in the mid-afternoon – rates are high because of air conditioning loads. Unsophisticated and vulnerable customers might have little option but to use their air conditioning during high-tariff times of the day and might end up paying more than they currently do. SMUD is keen to reduce peak electricity demand but do so in a way that protects vulnerable and poorer households. Many customers have voluntarily applied for the Peak Corps tariff that uses transponder technology to remotely switch off air conditioning for a proportion of an hour when demand is high. Consumers are paid when their air conditioning is switched off temporarily.

SMUD is well regarded by its customers. In 2010 it was the second most popular utility in the west region after Arizona's public utility Salt River Project. The top investor-owned utility, coming third, was Oregon's Portland General Electric.[14] Yola county, which adjoins Sacramento, has sought to integrate into the SMUD service area and leave the existing incumbent but this move has been resisted by the incumbent investor-owned utility.

not forthcoming, this precious investment lies underutilized and more importantly without the revenue stream to compensate the financial backers. Every quarter, nervous bondholders, shareholders and energy analysts will angrily demand why revenue is falling off.

The fact that private utilities that are obliged to maximize returns to share-holders are unenthusiastic about energy efficiency and some renewables agenda is well understood. Regulators have had to force energy efficiency and renewables from the investor-owned companies through new mandates. In 1982 California's energy regulator devised a cunning plan to decouple the profitability of the elec-tricity business from sales. But this is a far cry from making them want to actively improve energy efficiency.

California also mandates the investor-owned utilities to install energy efficien-cy measures, imposing penalties for significant underperformance and rewards that overcompensate the utility for achieving its targets. The regulator imposes incentives on consumers too: they are penalized for using more electricity than is regarded as necessary for their climate (the interior of California gets very hot in summer so households need to use more air conditioning) so the price of electri-city rises the more the consumer buys (so-called rising block tariffs). Box 5.4 gives a simplified account of some of the main programmes; each is much more com-plicated than suggested (for instance the tiered tariff has special allowances for households with special medical needs, who own an electrical vehicle or whose home is on a time-of-use tariff).

California has the most complex regulations and arguably the most sophis-ticated energy regulator in the world. But does all this work? The answer is a qualified yes. Use of electricity per person has been stable for around 30 years, while across the other 49 states (many of which also have utility energy efficiency programmes) there has been a 40 per cent growth. But critics of the programme, such as Mark Toney from California's customer representation group TURN, argues that much of the undoubted stabilization of Californian electricity use per person has arisen for reasons other than the utilities' energy efficiency projects, such as increasingly stringent minimum building and appliance standards, the shrinkage of California's (energy hungry) industrial base and favourable recent weather con-ditions. This reading of the data is disputed by many commentators,[15] but there is sophisticated econometric analysis to back up Toney's views.[16] He also argues that the regulatory system still incentivizes utilities to argue the need for new in-vestment in distribution wires and generation capacity, as the regulator will allow them a return on the capital invested. However, investor-owned companies are discouraged from spending their capital on their customers' homes (for instance in fixing the air conditioning) because the regulator does not treat this as capital that belongs to the shareholder that needs to be compensated.

The complexity of the regulatory system makes it opaque to everyone but the specialists involved in the process. Huge amounts of resources are spent on lobbying to influence key rules, on reporting on achievement of targets and their verification. The rulebook for evaluating, measuring and verifying energy

Box 5.4 Idiot's guide to Californian regulator's incentives for energy efficiency and renewables (much simplified)

Since California's nightmare experience of deregulation in 2001 it has reregulated the activities of its four investor-owned utilities: PG&E in northern California, Southern California Edison, San Diego Gas & Electric and Southern California Gas in southern California. Here are some of the energy efficiency and renewables regulations in operation:

- **Decoupling plus**: the scheme, strictly speaking called the 'Electric Revenue Adjustment Mechanism', allows a regulated energy company to increase prices if sales come in below the regulator's original forecasts. So for instance, if the allowed electricity price was 10 cents/kWh based on sales of a 100 units of electricity and sales fall to 90 units, the energy company's revenues would be insufficient to pay for the fixed costs such as maintaining the wires and networks. (The regulator does not need to compensate the utility for non-fixed costs such as fuel since the utility is spared the cost of buying as much gas and coal). The decoupling mechanism seeks to stop conventional power companies from losing revenue from energy efficiency. The programme is called decoupling plus because further incentives are applied to overcompensate achievements of targets, while there are penalties for underachievement of targets ('risk–reward incentive mechanism').[17]
- Demand response: regulators have asked companies to regard energy efficiency as the first priority in terms of preparing for the future. Companies are set targets to reduce the amounts of gas and electricity demand and also peak electricity demand. The utilities are expected to spend $3.1 billion between 2010 and 2012 and use this to help homeowners pay for low-energy lighting, heating, air conditioning and ventilation measures and subsidize the purchase of low-energy appliances.
- Tiered tariffs: the companies are required to offer customers tiered tariffs so the first units of consumption are at a lower price than subsequent units. Different 'baselines' are set for different climates within California to compensate for differing energy needs. Utilities set prices for four (or five) separate blocks. The first block called the 'baseline' is intended to be an adequate allowance to meet basic energy needs for a climate zone. It equates to 50–70 per cent of average usage. PG&E charge this at 12 cents/kWh, while consumption of the next block (between the baseline and 30 per cent more than the baseline) is charged at 13.5 cents/kWh. Charges for the third block (up to twice the baseline) are 29 cents/kW; anything beyond costs 40 cents/kWh.
- Renewable portfolio standard: since 2002, investor-owned utilities have been required to increase the share of renewables in their generation by 1 per cent per year and achieve 20 per cent by 2020. This has been met from wind, geothermal, small hydro (large hydro has already been exploited), biomass and some solar photovoltaics.

efficiency savings is 93 pages long. The budget for checking that rules are being followed is 4 per cent of the energy efficiency budget – around $125 million split between the regulator and the investor-owned company. The complexity and cost of engaging in this system act as a barrier to entry, deterring new entrants and small-scale solutions.

The most recent verification exercise carried out by the California Energy Commission,[18] which looks at both investor and publicly owned utilities, shows the investor-owned utility programmes are working well. The public utilities' performance in energy efficiency has been less ambitious and has achieved a smaller saving relative to the investor-owned utilities. But this might change. Until 2008 public utilities in California voluntarily undertook energy efficiency initiatives; after this date the state legislature introduced mandatory targets. So simply being a public agency *does not automatically mean it will invest in energy efficiency*. Public agencies have to be prodded too.

European countries (and other US states) generally have less sophisticated regulations than California. It is debatable whether smaller nations and states would have the critical mass of staff to develop and operate such smart regulation against large international energy companies.

Energy Service Companies

Instead of making utilities deny their own commercial instincts why not create businesses whose goal is to produce warmth and light as cheaply as possible? ESCOs are common in the commercial property market. They are the natural evolution of property management companies. The French firm Dalkia – which regards its core business as energy efficiency – is responsible for installing and maintaining heating, air conditioning and lighting needs of over 100,000 retail, commercial and industrial facilities, particularly in central and northern Europe. Unlike energy companies, their clients sign long-term contracts and pay for the type and level of service provided, rather than the fuel or power used. This gives the ESCO the freedom to invest in energy management (controls and timers to automate lighting and heating, sensors to detect if rooms are occupied), energy efficiency and ventilation systems (insulation of the property to reduce losses of heat through the envelope and heat recovery from stale air) and CHP and biomass in place of fossil fuels. Most of the energy utilities also have CHP businesses chiefly serving industrial customers. Many have attempted DH systems, operating as independent profit centres within the firm, but they often got their fingers burnt following the fall of gas prices in the late 1990s and early 2000s, coupled with the deregulation of gas and electricity markets, both of which hurt their economics.

Dalkia also operates 800 heat networks for local governments and industrial premises. A further 5 million households receive hot water from Dalkia-run DH networks. This all seems a lot – but the provision of centralized heat to households remains rare in the UK and North America and is only truly common in Denmark.

There are small DH schemes in the UK, and a handful in the US and Canada too. In Scotland, Aberdeen Heat & Power provides heat and power to 1000 households in tower blocks. It was set up in 2002 by the city as an autonomous not-for-profit business, using a private sector partner to design and build its projects. Like many other ESCOs established by local authorities in the UK, its purpose was to reduce the cost of energy to clients and work for the benefit of the citizens. The ESCO arranged the finance – through a mixture of borrowing, central government incentives and its own resources – and installed around 1.5MW capacity in CHP plants located in or close to buildings. The heat produced by the CHP is not metered but paid for through a fixed bill (heat meters might be retrofitted at a later date). This actually made the project easier to finance since banks found the stable and predictable stream of money easier to model. It also allowed the scheme to borrow cheaply, using the council's credit rating, which was superior to the tenants'. The tower blocks previously used expensive and carbon-intensive electricity for heating so there were substantial carbon and cost savings as expensive electricity was replaced by cheaper gas.

Wouldn't the lack of heat meters just encourage homeowners to heat their homes unnecessarily? 'This was a theoretical rather than an actual concern', said Michael King, the chairman of the ESCO.[19] The oft-repeated claim that households on unmetered energy waste energy by overheating their homes or leaving their windows open is not true in his experience. If they are provided with proper controls, most people will manage the temperature sensibly. The main objective of the scheme was to reduce the cost of energy and alleviate some of the ill health and discomfort caused by fuel poverty. Before the activities of the ESCO, most of the occupants were under-heating their homes, so it was quite proper that the internal temperatures would rise somewhat. The residents actually prefer a flat rate charge – they feel less inhibited about heating their homes to a comfortable level. They were also concerned that charges would vary unfairly in different flats – homeowners with greater external walls or on the ground floor or top floor would pay much more than a middle-floor flat. King's view that people living in a flat prefer to pay a flat rate charge for heat, rather than paying according to metered usage, is not shared throughout the DH or social housing movement. Others we have spoken to find off-putting the idea of people who heat their homes parsimoniously cross-subsidizing others who are, in their view, wasteful.

Cofely (a subsidiary of the large privately owned utility GDF Suez) runs a community ESCO in Southampton with a scheme serving commercial clients and households. It has invested in 50km of highly insulated DH heating pipes and also utilizes a renewable heat source: a geothermal borehole excavated in the 1980s. The scheme outputs the equivalent amount of heat as would be consumed by 30,000 homes. Many of the customers are commercial premises, so there is less of an overt social objective to the scheme than in Aberdeen.

A small but highly innovative scheme in Woking has a number of novel features that look at the energy needs of central Woking in a systemic way. The town's officers, Ray Morgan and Allan Jones, worked with Danish technical partners to form the ESCO Thameswey, which has developed a highly integrated scheme that uses 13 CHP units, DH and revised planning guidelines on all new developments, which requires new buildings near the centre of town to connect to the heat network. The council also installed photovoltaic panels using grants from the European Commission and the UK national government.

All of the UK schemes are handicapped by the unfavourable prices that are paid for small amounts of electricity, which the CHP facility exports onto the local grid, except if the facility uses renewable fuel sources such biomass, where it qualifies for feed-in tariff. The UK electricity market has been designed around the needs of large power stations and large centralized buyers of power. A small generator can often find it difficult to find a buyer for its exports onto the grid. Woking ended up installing its own private wires to link its disparate council buildings rather than pay the local electricity distribution company the charge to convey the power short distances.

Why haven't ESCOs offering DH and energy efficiency taken off in the residential sector? Michael King explains that the commercial environment makes it very difficult for companies to make money from an ESCO model in the UK:

> The difficulty in raising investment is the primary reason DH networks haven't taken off in the UK. The high up-front capital cost and low returns are not attractive to private investors. As well as this, ESCOs' revenues are exposed to risks from fluctuating gas price. Our customers want the price of heat to be the same as or cheaper than gas. But it's more expensive for us to insure against volatility in gas price using the futures market than for the gas suppliers. A big integrated utility which both produces and retails gas and power, isn't so exposed to fluctuations. When gas prices are high, one business division loses but another business division gains so the company as a whole faces little risk. Also it's very expensive to raise finance to fund the huge capital costs of new district heating pipes. Previous generations have already paid the capital costs for gas and power networks. Any new gas or electricity network is cheap to finance because the regulator Ofgem guarantees investors it'll provide them a return on their investment through regulated price increases. Also ESCOs don't get a good price for any electricity they sell onto the system … the existing electricity companies don't want to deal with us and we have to use third parties 'consolidators' to sell into the market which also charge a commission. The facts that combined heat and power is around 30 per cent more efficient at recovering power and heat from gas than the competition, and that each house is spared the cost and nuisance of looking after their own boiler, pale into insignificance.[20]

Energy Efficiency Utilities

An intermediate type of organization between an integrated utility and an ESCO is an energy efficiency utility. This is an organization whose mission is to improve energy efficiency but which is funded either from taxes or a levy on energy bills – essentially spinning off the energy efficiency obligations that are presently mandated on many energy utilities in the US, Canada and the UK. Efficiency Vermont (see Box 5.5) is an energy efficiency utility operated by the not-for-profit Vermont Energy Investment Corporation (VEIC) under a contract awarded by the state's utility regulator. VEIC is 'dedicated to reducing the economic, social, and environmental costs of energy consumption through cost-effective energy efficiency and renewable energy technologies'.[21] Blair Hamilton is the slim, silver-haired co-founder of Efficiency Vermont. He has been in the energy-saving business for 30 years. He explains that there is a fundamental difference between a utility that is mandated to install energy efficiency and a freestanding agency such as Efficiency Vermont: 'We have a somewhat unspoken advantage – there is alignment between the organizational and the public interest. Efficiency Vermont wants to save carbon. So there is less need for the regulator to look over its shoulder'.[22] That might be true in theory but it does not deter the state regulator from imposing substantial paper-based reporting requirements on the energy efficiency utility. The technologies deployed by Efficiency Vermont are similar to those used by other utilities: compact fluorescent light bulbs (CFLs), energy efficient appliances and HVAC systems. The straitjacket of verifiable emissions reduction programmes inhibits Efficiency Vermont from inventive behavioural or community-based decentralized energy because the reductions are often hard to verify.

In Hamilton's view there are some advantages to taking this role away from utilities:

> In California the utilities are doing a good job at rolling out energy efficiency – but it costs the rate payer more. The utilities ask for a 10–15 per cent fee (which is the rate of return they expect on capital investment projects), while Efficiency Vermont only receives a 2 per cent fee and Oregon no fee. Utilities are also more prone to gaming the system or ignoring it altogether. The reason why Efficiency Vermont was given the role by Vermont's Public Service Board was because several of the existing utilities didn't see energy efficiency as their core business and they kept missing their targets.[23]

Utility regulators in other states agree: Oregon, Wisconsin, Nova Scotia and Maine also have energy efficiency utilities.

Box 5.5 Efficiency Vermont – an energy efficiency utility

Efficiency Vermont has been awarded responsibility for delivering the state's energy efficiency targets. Savings of electricity from energy efficiency were 1.8 per cent in 2007, the highest rate of efficiency gain in any state – and the only state to outperform the target set by the state utility commission. As a result of this aggressive action, the state forecasts there will be no growth in energy demand between now and 2015.

Efficiency Vermont is tasked with reducing electricity use and peak electricity demand in all of Vermont (except for Burlington), taking over a function performed by the electric utility in most states. In the residential sector Efficiency Vermont saves electricity through giving away CFLs, incentivizing the purchase of Energy Star appliances, encouraging new homes to be built above building code standards in terms of airtightness, windows and heating systems. These measures are similar to the programmes run by conventional energy efficiency programmes run by utilities. Efficiency Vermont has voluntarily adopted an unfunded mandate to reduce gas consumption at no cost to the ratepayer, in order to fulfil it broader mission. Efficiency Vermont staff cajole architects and building professionals into incorporating zero-carbon building techniques, such as orienting buildings to maximize passive heating gain, into the design.

The cost of running Efficiency Vermont adds $35 to each Vermont customer's bill – again the highest in the country. But its results are impressive. The price of electricity in Vermont is $0.13/kWh but the cost of investment per unit of first year saving was $0.22/kWh, suggesting a payback of less than two years[24] and $0.02 per kWh across the lifetime of the technologies.

Drawbacks of Existing Energy Efficiency Organizations

I installed a hot water solar panel a few years ago, I asked for a small photovoltaic panel to provide power to drive the pump, I suggested a spot where there was loads of sun. But he installed the photovoltaic panel in the corner, which was shaded by the chimneys in the morning and late afternoon. We'd only get hot water from the panel for a couple of hours as there wasn't enough sunshine on the photovoltaic panel to drive the pump. In the end we had to connect the pump to the house's power circuit. It makes me mad he just ignored my advice.

A friend's experience of fitting a solar hot water system

I asked the contractor to provide some evidence the double-glazed windows had been filled with argon as I'd requested and he hadn't just fobbed me off with air. In my email I asked for 'Ar' – I think he thought it was just a spelling mistake. He lifted the windows and gave them a shake: 'Can you hear that rattle? It's the sound of argon', he explained sagely.

Prashant's direct experience of a contractor

Even if an organization's commercial interests and motivation are perfectly aligned with the repowering communities agenda, this is not enough to deliver energy efficiency and local heat and power generation. This should come as no surprise. Retrofitting old homes to dramatically improve their energy efficiency is a substantial and disruptive job, which affects the aesthetics, market value and liveability of the home. These are not commodity goods; they are highly bespoke, requiring skill and training. Local energy production, for instance using waste incineration or biomass, is also disruptive and requires participation in community decision-making. We are asking for a huge switch: from supplying ever greater quantities of energy to changing the fabric of communities so they need less fuel and so they derive greater useful energy from the fuel that is burnt.

For a business to have credibility with households that it can deliver the promised benefits, it should have a track record of undertaking these sorts of projects. It needs to be long lived and deep pocketed enough that customers believe it will be there in the years to come when problems might surface. It needs to have excellent listening skills to ensure that worries, both said and unsaid, are taken on board. It needs to have the skills to select (or train) the different trades and professionals. It needs to have systems to control the quality of the work, including the unseen work that has to be done well to avoid hurting the energy efficiency of the system. These sorts of organization do not exist yet. So we need innovation in organizations as well as in technologies.

We've all had our ghastly experiences when the construction industry is installing something it understands well. It's even worse when the contractor is installing something he has little experience of. Ideas such as thermal bridging and airtightness have to be not just learnt in a textbook sense but become second nature. A single metal nail or pipe that passes all the way through an insulated wall can greatly hurt the performance of the wall.

And it's not just the supply chain. Government agencies need to retrain to understand too. An architect friend of ours tells of a house that was expensively insulated with external insulation on its side walls. But then the planning authority insisted that insulation could not be externally applied to the chimney breasts (which protruded a few inches from the rest of the wall), which would have made the side wall jut out another few inches. For 100 years, the town's citizens had gone about their day-to-day lives, despite the fact the chimney breasts had stuck out by a few inches from the rest of the house. But now the planning authorities could use their powers to kill off this unsightly protuberance, which they did with gusto. But as a result cold from the outside could penetrate the brickwork and spread sideways through the wall, undermining the effectiveness of the insulation.

Blair Hamilton at Efficiency Vermont is surprised at the lack of priority that official energy efficiency programmes in the UK give to improving airtightness in energy efficiency retrofit. In the US and Canada, retrofit projects start by making the building as airtight as possible and introducing controlled mechanical ventilation in areas where needed. In the UK, the average house exchanges its air around 11 times an hour; this is 14 times the standard allowed in Sweden – admittedly the world leader in energy efficiency practices. The measured airtightness in the UK is only about a third of that in Canada, Switzerland and Sweden. There is the perception that newer homes have been built to better standards, but the data show that at least until 1994 this was not the case and airtightness had not improved through a century of construction practices.[25] The leakiness of UK homes is a major issue and a cheap one to rectify. In the unofficial community-led energy efficiency movement undertaken by local authorities, Transition Towns and barn-raising parties in the US, draught-stripping is the first measure fitted by energy efficiency enthusiasts.

Why do official, industry-organized programmes not undertake improvements in airtightness? The biggest energy efficiency programme in the UK run by British Gas offers just cavity wall and loft insulation as standard, because this is what the company is incentivized to do. Airtightness is not credited because it is hard for the regulator to verify that the work has been done without expensive airtightness testing. The UK academic David Olivier is more scathing. Writing in 1999 he says:

> There are many reasons why UK air leakage figures are still so high. One clearly is that most designers and builders still believe that their current practices are wholly satisfactory, and are genuinely surprised when field tests show otherwise. Typically, there is a lack of sound design details, reasonable to rather poor site workmanship and total ignorance of the principles of airtightness on the part of site management, so supervision is sketchy or ineffective.[26]

Maybe things have improved since then – new homes are tested for airtightness – but even this apparent improvement has been compromised through deregulation of the regime, which allows builders to select the inspector, creating an obvious conflict of interest in the verification process.

The UK energy regulator's attitude is understandable. Making lofts, doors and windows airtight requires attention to detail, patience and respect for the character of the home. Paul Zabriskie who manages central Vermont's low-income insulation programme sums it up well: 'weatherization is all about skilled people continuing to do unpleasant work well when nobody is looking. For this you need people you can trust. Access to the majority of attics is in the main bedroom – this is the most intimate space in someone's home'.[27] His team is deeply bedded into the community: like most other weatherization programmes it has a twin mandate of improving energy efficiency and also providing jobs and skills to the community it serves. It is run on a not-for-profit basis. His organization, in partnership with Efficiency Vermont, has produced an excellent how-to educational video called 'Button-up Video'.[28]

There is a widely held fear that we need to rely on large energy companies to deliver energy efficiency, that only they have the size and talent to roll-out energy efficiency at scale. Many of the case studies regularly cited are dismissed as wilful individuals in government battling against the odds. We think the truth is actually

Box 5.6 Islington – a typical London borough

Many people with whom Stephen and Prashant have discussed *Repowering Communities* have questioned whether local government is capable of playing a pivotal role in delivering community energy. The widespread view is that UK local government does not have enough power or money to attract talent. Stephen asked his council to show him around the social houses they manage. In the UK social housing has better energy efficiency performance than privately owned property. This is exactly the opposite of the so-called landlord–tenant paradox we hear about. Mandatory standards imposed on social landlords mean social rented accommodation is better than owner-occupied and privately tenanted properties.

Islington is a part of London with some very wealthy inhabitants, but also many low-income inhabitants, so it has many council homes. It is a large borough with a population of 190,000. It is not known for its record on sustainable energy issues. If anything it has a reputation for being a poorly run borough. As recently as 2002 it was considered by its auditors as being among the 25 per cent worst performing local authorities in England and as 'performing a poor service for its residents'. Its performance has since improved.

Stephen was surprised to find out that the council was quite active in improving the energy efficiency of the buildings it owns. It has clad some high-rise tower blocks with 10cm of external insulation. The blocks have also been fitted with efficient lighting in corridors and other common areas, reducing electricity use in these areas by around 65 per cent. Another block of flats had green roofs fitted that, as well as being good for wildlife, improved urban water drainage and is visually attractive. It is also excellent insulation.

A micro wind turbine has been fitted to the roof of a 17-storey block of flats – a good location for harnessing the wind. However, it cost about £70,000 to install, and Stephen's guide thought that there was limited scope for expanding wind power in the borough. More impressive were the solar thermal panels on the roof of another block. These provided enough hot water for 18 flats.

Much of the work was carried out using grants from central government, but in some cases the council used its own capital funds to top-up the grants. The capital funds were recouped by one-off charges to leaseholders. Residents have benefited not only from warmer homes, but also from having to pay less annually for the electricity used in communal areas and for maintenance and replacement of lighting equipment in communal areas.

The officer guiding Stephen was knowledgeable, articulate and determined. There are many people like him, quietly getting on with things around the country. Local government has few powers in the UK, budgets are under pressure, especially for discretionary activities that go beyond services local government is obliged to provide. Despite this there are good people working in councils already and, if local government is given greater powers, more will be attracted.

that local governments often do good work but this goes unnoticed in the national psyche and sometimes within their own communities. Local government and social landlords are substantial owners of houses and on a day-to-day basis manage and invest in their properties. Box 5.6 discusses some actions Stephen's home borough has taken.

Improving the Skills of Energy Efficiency Workers

How do we develop the necessary skills among contractors and make customers more savvy? In successful schemes it is common for local government to show leadership by insisting on high construction standards and distributed energy in the municipal buildings and social homes it controls. For example, SMUD maintains a qualified contractors list. To get on the list, a contractor must be properly licensed and offer a minimum of a two-year warranty. Before contractors can give advice under SMUD's Home Performance Program, they need to go through training of three to nine days. SMUD has a good reputation among its customers, and contractors that have been through the training benefit from this shared glory. According to one of the executives at SMUD, trained contractors act as a marketing arm for energy efficiency and the SMUD training programme.

Berkeley in California has implemented the Residential Energy Conservation Ordinance (RECO). This requires homeowners that are selling or undertaking major refurbishments on their homes to install basic water and energy efficiency measures, including aerated shower heads, low-flow toilets, water pipe insulation, water tank insulation and insulation of heating ducts. Together these cost $1200–2000. This policy is being applied to 500–700 homes – around 0.25 per cent of Berkeley's stock of homes each year. Few other places have mandated compulsory energy efficiency retrofit. Billy Romain explains: 'People feel comfortable with the RECO. Its part of being of being a Berkeley person'.[29] But the costs of providing audit (including energy modelling) and verification of the scheme are high at $1000 per treated home.

Efficiency Vermont has intervened to raise skills and knowledge of the local building industry. Contractors that undertake work for Efficiency Vermont have to attend and pay for training courses organized by Efficiency Vermont – they receive part of the training costs back once they pass the course and complete two paid-for jobs. Blair Hamilton argues that the design of the training policy encourages commitment from the contractors.

In the UK, British Gas has invested in training facilities to improve the skills and knowledge of its frontline staff, who visit people's homes and advise clients about energy efficiency. The City of Toronto's Energy Efficiency Office is responsible for the management of the Better Buildings Partnership that promotes and implements energy efficiency and building renewal in the municipal, academic, social and healthcare services sector, not-for-profit sector and, most recently, privately owned multi-family residential buildings. The Better Buildings Partnership was created as a programme to address climate change and its implications for society, economic development and job creation. These three areas are central to a number of objectives that the city council mandated for the partnership at the time of its inception in 1995. To date, the Better Buildings Partnership has retrofitted nearly 1500 buildings, saved 1.8 million tonnes of CO_2 and has created 16,900 person years of work. Typical measures including improving building envelopes (windows, air sealing and insulation), HVAC upgrades and building automation/controls and lighting upgrades.

The Better Buildings Partnership staff provide technical assistance in audits, review feasibility studies and approve applications. The programme works by providing interest-free loans for energy efficiency and renewable energy projects to eligible building owners for up to 49 per cent of the project value and it assists in identifying other grants available from the city, provincial and federal governments. The typical payback period for projects is seven to ten years and typical loans are paid off in 10 to 20 years, regardless of the magnitude of the energy cost savings or energy generation realized by the project.

Audrey Schulman runs a non-profit Home Energy Efficiency Team (HEET) in Cambridge, Massachusetts. HEET has installed energy efficiency and motivated behavioural change from residents for a fraction of the cost of the utility programmes but failed to obtain any material support from utilities, despite attempts to engage with them. Volunteers donate three hours of their time and install basic energy efficiency measures to others in their community. Schulman states:

> Because we are a grassroots non-profit, people trust us and our recommendations. We are part of the community and not motivated by profit. Over three quarters of the residents whose homes I evaluate end up hosting an energy-upgrade party. We ask the resident to pay for materials and we typically reduce gas and both heat and electricity power consumption by 10 per cent through air sealing, installing basic insulation, fitting CFLs and the water and electrical efficiency measures, getting a return on investment of two years or less. On the other hand, many people are suspicious of the motives of the utility companies and the contractors. The contractors sent out by the utility companies as part of their energy efficiency program are only able to sell a third of the residents actual installations even though with the rebates the cost would be very low.[30]

Bringing a Community Together

> I have certainly never been eco-friendly, energy conscious in any way at all ... But once we got involved with it, we realised how much it could save us, firstly. But the implications for, in a wider sense of having a two-year old daughter, possible effects to save CO_2 emissions, you start being more conscious of it, reading the newspapers, on television and whatever.

Participant in British Gas's Green Street pilot project, Edinburgh[31]

Installing internal energy efficiency measures like draught proofing or new heating systems does not affect others in the community. It makes sense to insert underfloor insulation when the house is being rewired, or insulate the loft when it is next emptied.

But changes to the outside fabric of the building such as applying external insulation, changing windows, or installing solar hot water panels can benefit from economies of scale through adopting a street-by-street approach, or a whole-building approach in a block of flats. In streets with heritage value there might be an advantage in ensuring that changes to the look of a home is coordinated across a neighbourhood. When it comes to installing DH measures, the economics rely on homes linking up to the heat network soon after it is installed. The developer has to pay interest on the capital invested and the longer the network's capacity goes unused, the worse the economics of the project. And perhaps most importantly, and as suggested in the quote above, acting together at the community level engages and enthuses people to take actions to reduce their energy use, which they feel powerless to do now.

How do organizations go about organizing street- or community-scale action? There has to be a compelling reason for people wanting to do it. Some of these changes are disruptive and a nuisance to organize and live through. The motivations are varied: saving money, improving the comfort of the house, fitting-in with others and concern about climate change. The factors that hinder community action include: variation in people's capacity to pay, lack of time or inclination to be involved, mixture of tenures in the property, difficulties and cost of reaching decisions or developing a consensus.

The process of negotiating and securing a community project is important. Who could best lead it – an energy company or ESCO, the local government, the owner of a housing block? What is the role of local residents – to act as sounding board, to act as ambassador, to run the process? What degree of consensus is needed before a project can go forward: unanimity with no one opposing the proposal, a large majority agreeing to it, or a simple majority agreeing to it? And how should the views of the disengaged be incorporated: as passive agreement, indifference or opposition? Sometimes the residents' association may not function effectively and may rarely meet.

The UK established a Community Energy Programme for a number of years to help the development of DH schemes. Around £50 million was allocated but few projects were completed because money had to be spent within the fixed timeframe and the process of negotiation and completion of work did not occur fast enough. The UK has launched the Community Energy Saving Programme, which requires energy companies to undertake community energy efficiency actions in some of the poorest neighbourhoods of the UK. Again there are fixed time limits for the energy companies to complete the work. Community action takes time to overcome the various barriers described above. But as the evaluation of British Gas's Green Street shows,[32] the ability to use neighbourliness and community spirit means these programmes can reach people that would otherwise never consciously think about their energy use.

Aberdeen Heat & Power overcame some of these issues and managed to secure funding for part of their costs. The city council owns and manages the external structure of the building. As is common in the UK, the residential blocks have mixed tenure: some are owned by a landlord and others by the residents (leasehold). Leaseholders had to pay for their share of the costs for any communal investments, such as external walls or new heating systems. Because many of the residents – both tenants and leaseholders – were on low income, the scheme qualified for support from the energy company owing to its obligation to insulate the homes of so-called priority groups. The city had to raise some of the finance from the banks. Because many of the leaseholders had a poor credit history, they would have struggled to borrow money themselves, and indeed would have adversely affected the city's borrowing application. The council successfully applied for grants to cover most of the leaseholders costs and the balance of £1500 per home was paid by the leaseholders themselves.

Efficiency Vermont is required to target a proportion of its energy efficiency measures in particular geographic areas where the electricity distribution system is running close to capacity. This targeting of their efforts allows the energy companies to avoid expensive work to upgrade the wires in that area and can significantly reduce capital spend. It means that the system as a whole can work more efficiently. It can also be used to target areas where the climate is harsh and energy use per person high. Trying to step up the penetration of energy efficiency in areas where the grid is near capacity was, as Blair Hamilton puts it:

> Difficult, expensive and great fun. We go about it by working through town energy committees and helping them organise an event 'Can Poultney light the way for Vermont?' was our slogan. We arranged an event to coincide with a holiday weekend. We worked with the local hardware store to ensure there were enough bulbs. The offer to entice customers was one free CFL and up to six extra for just 89 cents each. On average people took home four bulbs each. We got a list with the names of all the people in the village from the utility and over the weekend went door to door persuading everyone to take at least one bulb. At the end of the day there was just one name left on the list – a friend of the mayor's and he personally went and delivered the bulb.[33]

There is a limit to what can be achieved from bottom-up processes and consensual development of low-carbon initiatives. In particular, building DH heating systems into existing communities can be economically positive (if low public sector interest rates are used), but will often take subsidies to break even. Public bodies have to lead the community and assess whether such networks are the best system for delivering heat in *anticipation* of future availability, the price of gas and carbon targets.

Denmark has shown such strategic leadership and over the past 30 years small-scale DH networks have been linked with heat transmission pipes, coal has been replaced with lower-carbon gas CHP, and more recently with low-carbon energy from waste combustion and biomass. In the future it is possible that waste heat from fossil fuels plants with CCS could also be used. These are large-scale strategic decisions that have to be made at the national and provincial level, and then implemented at the community level. In Denmark the Heat Law created a situation where local authorities in urban areas were given the mandate to *require* householders and commercial premises to join DH networks within ten years of the network becoming built. This forced homeowners to scrap their old heating systems.

There is no single approach to these issues and in most countries collective action to improve energy efficiency or install local heat and power is the exception rather than the norm.

Energy companies have a track record at reducing energy use in households in North America and the UK. Without doubt they employ staff that are expert in their fields and sincere in their efforts to do their jobs well. Yet this chapter argues that we need to move away from this approach and strip away these duties from these companies. Are we just being dogmatically anti-capitalist?

We have spoken to senior staff in utilities in both North America and the UK who say that they have a difficult job going to their investment boards making the case for expansion in energy efficiency or local energy. The idea they are pitching – an idea that would save the customers money and shrink the market – gets a tough ride. Even if as human beings the board members might agree with the proposal, as board members they tend not to.

We believe that the current model of existing integrated energy companies having significant responsibilities for delivering reductions in energy use or distributed production is flawed. The challenge ahead of us is not selling low-energy commodity goods such as appliances or light bulbs (though we have to do these too), nor commodity energy efficiency measures such as cavity wall insulation. The challenge is to promote complex bespoke projects that have to take into account and influence customer and community attitudes.

We asked a utility director in the water sector why his company was not more involved in customer-side plumbing repairs, or work in homeowners' properties to reduce water consumption (in the UK water companies do not have compulsory water efficiency targets as energy companies do). His company employed many engineers and plumbers, the brand was well known and respected, and they already had many millions of customers. He shuddered at the prospect. The idea of his company having to negotiate and sometimes take unfair criticism for badly finished plumbing jobs made him fear for his company's reputation.

Deciding what energy infrastructure should be built in a territory is an intensely political choice. It has to balance competing interests, help local economic needs, especially for employment, and local commercial interests. But there needs to be a financial model that allows energy saving or local energy production to be a viable business proposition. We are attracted to the ESCO model. But for these to prosper within a liberalized energy market, such as that in northwest Europe, there have to be changes to insulate these companies from the rest of the energy market. Only local government has the political authority to make these difficult choices.

These plans should be delivered by municipal-owned bodies, not-for-profits or co-ops that have greater legitimacy with the local population. The appropriate organism would depend on the tasks that the local authority has set. These might have to include social housing organizations, local voluntary groups or in rural areas customer-owned co-ops. Beyond a certain size there would have to be more bespoke organizations solely responsible for delivering large infrastructure projects or managing networks. Rather than creating competition at the level of retail provision, competition could occur when the contract is first offered.

Home retrofits, especially of older homes, require the contractor to tailor their service to individual homes, have sensitivity to community-level politics, to navigate aesthetic mores, and have excellent inspection and verification procedures. The companies are not the first port of call when people need minor construction works or maintenance, but these will usually be the best triggers for installing insulation.

The creation of new heat networks establishes many new local monopolies. It will never be feasible to regulate large numbers of such entities. Instead these must have the fair treatment of customers written into their constitutions. These will need to be owned by communities and not for profit to provide people with confidence that they won't abuse their monopoly powers.

Notes

1   Blair, T. (2010) *A Journey*, Hutchinson, London.
2   See www.famous-quotes.net/Quote.aspx?No_institution_can_survive_needs_ geniuses_manage.
3   Co-operatives UK (2010) *The UK Co-operative Economy 2010: A Review of Co-operative Enterprises*, Co-operatives UK.
4   DTI (2004) 'Cooperative energy: lessons from Denmark and Sweden', Global Watch Mission Report, UK Department of Trade and Industry.
5   Danish Wind Turbine Owners' Association (2009) 'Cooperatives – a local and democratic ownership to wind turbines', www.dkvind.dk/eng/faq/cooperatives.pdf.
6   Baywind website, www.baywind.co.uk/baywind_aboutus.asp.
7   Dresner, S. (2010) 'Gigha case study', http://climateanswers.info/2010/06/ repowering-communities-gigha-case-study.
8   Highlands and Islands Community Energy Company (2008) 'The electrification of Eigg', www.communityenergyscotland.org.uk/userfiles/file/Case20studies/Isle20of 20Eigg20electrification20case20study.pdf.
9   News Review (2005) 'An employee working for the PR agency Meridian Pacific to hack into SSJID's data during a public meeting', www.newsreview.com/sacramento/ content?oid=44039. PG&E settled for $404,000 following an FBI investigation.
10 American Public Power Association (2009) *2008–09 Annual Directory & Statistical Report*, www.appanet.org/files/PDFs/ppcostsless2006.pdf. Original data from the

Energy Information Administration. Data were reported by 2010 publicly owned electric utilities, 217 investor-owned utilities and 882 co-operative utilities operating in the 50 states and the District of Columbia.

11 LAWP (2009) www.ladwpneighborhoodnews.com/external/content/document/1643/280504/1/LADWP%20Strategic%20Planning.pdf.

12 Edison International (2009) *Edison International Annual Report*, www.annualreports.com/HostedData/AnnualReports/PDFArchive/eix2009.pdf.

13 See www.smud.org/en/board/Pages/strategic-direction.aspx.

14 J. D. Power and Associates (2010) *Electric Utility Residential Customer Satisfaction Study*[SM], http://businesscenter.jdpower.com/JDPAContent/CorpComm/News/content/Releases/pdf/2010120-eurs.pdf.

15 National Resources Defense Fund (2010) 'California restores its energy efficiency leadership', NRDF, San Francisco, CA.

16 Mitchell, C. (2009) 'Stabilizing California's demand - the real reason behind the state's energy savings,' *Public Utilities Fortnightly*, www.fortnightly.com.

17 California Public Utilities Commission (2010) *2006–2008 Energy Division Scenario Analysis Report*, www.ftp.cpuc.ca.gov/gopher-data/energy%20efficiency/Final%20Energy%20Division%20Scenario%20Analysis%20Report_070910.pdf.

18 California Energy Commission (2009) *Achieving Cost-Effective Energy Efficiency for California: 2008 Progress Report*, California Energy Commission.

19 Prashant's interview with Michael King, 11 November 2010.

20 Prashant's interview with Michael King, 11 November 2010.

21 See www.veic.org/index.aspx.

22 Prashant's interview with Blair Hamilton, 5 June 2010.

23 Prashant's interview with Blair Hamilton, 5 June 2010.

24 Navigant Consulting (2008) *Benchmarking of Vermont's 2008 Electric Energy Efficiency Programs: A Comparative Review of Efficiency Vermont and Burlington Electric Department*, http://publicservice.vermont.gov/energy/ee_files/efficiency/Final%20VT%20BED%20Benchmarking%20Report.pdf.

25 Olivier, D. (1999) 'Air leakage standards', www.communities.gov.uk/documents/planningandbuilding/pdf/133367.pdf.

26 Olivier, D. (1999) 'Air leakage standards', www.communities.gov.uk/documents/planningandbuilding/pdf/133367.pdf.

27 Interview with Paul Zabriskie, 2 June 2010.

28 This can be downloaded at www.cvcac.org/index.php/weatherization/button-up-vt.

29 Interview with Billy Romain, 18 June 2010.

30 Interview with Schulman 10 June 2010.

31 Cited in Platt, R. (2010) 'Green Streets: Exploring the potential of community energy projects', IPPR, www.ippr.org.uk/members/download.asp?f=%2Fecomm%2Ffiles%2FGreen+Streets+interim+report+web.pdf.

32 Institute of Public Policy Research (2010) 'Green Streets: Exploring the potential of community energy projects', www.greenstreets.co.uk/files/157.

33 Prashant's interview with Blair Hamilton, 5 June 2010.

Chapter 6

Financing Community Energy Initiatives

Why would I spend my grandchildren's hard-earned inheritance on a project like this?

Theo Paphitis[1]

In the widely syndicated show *Dragons' Den* (*Shark Tank* in the US) entrepreneurs pitch their new business ventures to four wealthy 'dragon' investors. If their pitch is successful the dragons take a stake in the business; the dragons are looking for the entrepreneur's proposition to make sense from a business point of view and the individual has to have credibility that he or she can deliver what is promised. Each week most of the budding businessmen and women leave empty-handed. Their pet ideas, which friends and family assure them will make them rich, fall apart under financial scrutiny. The reasons for failure are varied: over-optimistic sales forecasts, insufficient protection of intellectual property, little prospects for growth, lack of confidence that the venture will scale, or profit margins too slim. This is the world of finance. Grotesquely unsentimental, the dragons are interested in whether the idea will make sense in the world as it is, not in some theoretical spreadsheet, nor in the entrepreneurs' fanciful imaginings.

What would happen if we pitched community energy to the dragons? We'd have to be clear about the precise product we had in mind. The type of investment we are seeking finance for is community energy – by this we mean area-based energy efficiency programmes, new networks (such as heat pipework), small-scale renewables of up to 50MW (sufficient to power a small town) or CHP plant of up to 50MW (sufficient to power and heat a small town). These are the types of investment that engineers and economists favour when looking at the efficiency and cost-effectiveness of the system overall, but they are tricky to finance. We would be excluding large-scale power plants – these can raise finance relatively easily already. Investors understand these sorts of investments; they are safe bets and will make a return if well managed. We'd also be excluding investment in electricity and gas pipes that have been approved by the energy regulator. Again these are easy to finance since, once they are constructed, the regulator allows the energy company to pass on the cost to customers (see Box 6.1). We'd also be excluding minor energy efficiency measures in the home that cost less than €500 since these can usually be financed by the household or business.

If such energy efficiency and community generation ever did try to appear on the show, the sad fact is it'd never even make it past the auditions because the sum of money needed vastly exceeds the £200,0000 maximum allowed in the UK version of *Dragons' Den*. Community energy is expensive, costing far more than the sums a few business angels can cobble together. A 2010 report by the management consultants McKinsey & Company[2] puts the cost of enhancing the US's

Box 6.1 Raising funds to finance new electricity and gas transmission and distribution

The companies that own the electricity and gas transmission and distribution systems are monopolies. No customer has any choice but to use this infrastructure. No generator has any choice but to connect their plant to their systems. Governments have established economic regulators that set a charge for the use of the wires and pipes, or so-called regulated assets. This charge is based on the value of the assets and a reasonable rate of return to provide shareholders and banks with a profit. This charge is added to the cost of generating the electricity and producing the gas.

But how does the company that owns the wires and pipes get paid for any investment it makes in extending or strengthening its networks. The regulators have designed a cunning mechanism to reward new investment. First, the regulator asks the company what its investment and maintenance plans are for the next few years. It looks at this list sceptically, figuring that the company has every reason to exaggerate its spending need. Once the projects are agreed and the capital-spending plan approved, the company is allowed to seek finance and commence work. The approved spend is added to the regulated asset base (RAB), fully rewarding investors for their investment by allowing energy prices to rise. Borrowing money to finance approved investment is very easy since the customer is guaranteed to pay (they're cut off by their energy supplier if they don't) and so network companies typically have access to very cheap finance – cheaper even than some insolvent governments.

Use of the RAB is restricted to investment that the regulator approves of. It can't be used for assets that aren't approved, for instance heat networks or improvements to the energy efficiency of the building fabric. Such assets won't have access to cheap loans. Not unless local or central government guarantees these loans or itself lends the projects money itself, as happens in rural US, Denmark and Germany.

129 million residential homes at $229 billion (but values the current and future energy savings at $395 billion). The European Commission says €1 trillion has to be invested in energy systems by 2020 to hit energy security and climate change goals.[3] This report goes on to say that the opportunities for energy efficiency are mostly untapped and that the EU nations could save as much as €78 billion a year by 2020 in energy costs by installing cost-effective energy efficiency opportunities. Those returns are not being realized, partly because prospective investors feel uncertain about the policy and economic environment. In economists' eyes, this fear is irrational because the energy savings all cost less than the lowest current cost of investing in new power plant. Surely some sort of deal could be done between generators and their customers to avoid building expensive new plant?

The costs listed above, while substantial, still don't include the much higher costs of investments in improving the energy efficiency of existing buildings beyond the low-cost measures. This requires investment in measures such as solid wall insulation. How much such 'deep retrofits' cost is uncertain. A 2009 report by the UK energy efficiency industry[4] puts the cost of external solid wall insulation at £14,000 for the average home once the cost of scaffolding, extending the roof and refitting the guttering is taken into account. A recent estimate of the cost of deep retrofit for UK buildings put a figure of £230 billion for investment between 2010 and 2020.[5] This just covers the cost of upgrading the building shells and the heating and cooling equipment in existing homes. A community that seeks to introduce DH has to pay for long-lived pipework. The UK government estimated that £50 billion might be needed to install community DH networks for 5.5–6.5 million households.[6]

But let's pretend that a community energy project gets past the audition and gets on the show. Before the dragons declare: 'I'm out', what kind of deviancy would they accuse the project of?

In other chapters we speak about some of the behavioural reasons why people and businesses don't feel confident about spending money on energy efficiency (see Chapter 8), and the lack of financial reward for installing renewable or more secure forms of energy (see Chapter 7). These issues need to be remedied but there are other reasons for the dragons' reticence. For one thing there are costs associated with these technologies being unfamiliar (at least in some countries): installers are learning on the job and equipment providers are pricing their wares high because they haven't got economies of scale. Second, there is the landlord–tenant issue: the building owner, perhaps a commercial landlord or a social housing provider, who has to arrange the external insulation is different from the person that pays the energy bill and benefits from the improvement. Third, the community might have divergent views – not everyone in the community will want the investment to take place – some might even campaign against it and cause a stink in the press and delay the process, destroying already thin margins. Fourth, some people in the community who wish to take part might have poor credit records and could not be relied on to pay their bills. Lastly, the investment might not necessarily be remunerative at the high interest rates the dragons' banker will offer for this sort of project.

The last is a major impediment – but there is a degree of circularity in the problem. Because the technologies are relatively uncommon in most countries, they are seen as risky and become expensive to finance. Once they become more familiar, the unit costs will fall and lenders will be more used to them and stop requiring such a hefty return from their 'risky' investment.

Let's do a quick thought experiment to show how important sentiment and technology immaturity is. Suppose the cost of retrofitting a home is £10,000 per home. If the occupier succeeds in reducing his energy use by half – not an un-reasonable target for a deep retrofit – the savings will be worth around £600 per year at today's energy prices – a rate of return of around 6 per cent. So it will take almost 15 years to pay off the loan. No one has this amount of capital sitting in a zero interest bank account. People will need to borrow the money from a bank or dip into their savings – but at commercial rates of interest, which range from 5 per cent (if added to a house loan) to 15 per cent (if taken out as an unsecured loan), this investment looks at best like a marginal investment, and at worst a crazy investment. Far from being free money, deep retrofits only make sense to people who have ample savings sitting in a low-interest account.

But let's see what happens if we play about with some of the assumptions. The numbers above assume that the price of energy stays constant. As gas be-comes scarce or harder to get from convenient locations, the price of gas and electricity (in areas where the price of electricity is linked to the price of gas) will rise. In the US, the price of electricity is more likely to follow the price of coal. Also the cost of the insulation could come down if there are technological im-provements – Prashant has been to a demonstration where lasers were used to accurately measure the internal walls so insulation material could be cheaply cut off-site. There are also economies of scale if someone organizes the whole street or building to be insulated in one go.

If energy prices rose by 2.5 per cent a year, the project would generate a return of 8.5 per cent; this is acceptable to banks if the loan is added to the mortgage. If the cost of the investment is reduced – either through government subsidies, innovation or economies of scale – it looks an even better bet. Still not enough to set an investment banker's heart aflutter, but a tidy return. If the bank or energy company is certain it will be paid back, this is something that could in theory be financed. Sadly banks and businesses aren't certain. Investing in improving the energy efficiency of people's homes is presently seen as a risky investment. A working group of the UN's Environment Programme[7] concludes:

> *Energy savings ... are not a conventional 'asset' against which a bank will lend ... cash-flow from energy savings is not a familiar form of revenue or collateral to back lending ... financial institutions, particularly local ones, need to become familiar with the nature, as well as the performance and credit risks of energy savings financed projects in order to be comfortable with providing debt.*

Funding energy efficiency is difficult for banks for a number of reasons. First, the sums of money being lent – say around £10,000 per home – are small compared to the cost of setting the loan up. Second, what if the homeowner doesn't pay? The investor can't just peel the insulation off and sell it to someone else. Or what if the insulation doesn't perform as expected – will the investor be exposed to further costs in having the problem fixed, or be exposed to the customer's ire?

If anything, the situation is even worse for communal investments like DH. By their nature the economics of collective or community investments are tricky. There is the scope for substantial economies of scale: a 10MW CHP plant is far cheaper to build and operate than 1000 10kW micro-units. But the costs of linking a community by DH pipes can be very high. (However, once these costs are incurred and the network is established, it requires little maintenance and lasts for decades.) The worst worry for an investor is the uncertainty about whether people will choose to connect to the heating network. They will be aware of the many network companies that have gone bust, not because the business proposition or technologies were flawed but because the company had incurred huge debt building the water, fibre optic, railway (delete as appropriate, there are examples of bankruptcies in each of these sectors) networks, and then haemorrhaged cash in the first few years of trading. In these early years, the company has to pay interest to its creditors but it has a small customer base. The irony is that from a societal perspective the new network might one day become a huge success and, though the original investors who took the risk lose their shirts, future generations benefit from these long-lived assets.

In the Den, the dragons sit behind a big pile of money – a palpable demonstration of the gulf between those with capital and those wanting capital. Over the course of the short (and often humiliating) audience, the dragon and entrepreneur negotiate the terms of a possible deal, so that all parties know what is expected of them.

If money is to flow from financiers into community energy, investors aren't going to be wooed by dizzying financial returns. It's not that sort of business. The capital markets have to start viewing energy efficiency and distributed generation as a safe and low-risk investment – it has to be boringly reliable. Sorry dragons, we don't want you. Community energy has to flirt with pension funds, not venture capitalists. To do this, the policy framework has to be stable and predictable, and renewable and distributed energy has to be fully rewarded for its environmental and energy security benefits. This would be a dramatic reversal from the present situation. So the big question is: how can we make community energy a boring and predictable investment?

How Does Community Energy Get Financed at Present?

Financing community energy in real life has some similarities to the Den. Someone has to provide the finance and someone has to be the 'payee' responsible for paying back the investor. There also has to be a set of rules and conditions to govern how the money will be repaid and what happens if things don't go according to plan and someone has to picked up the pieces – we can call this the financing vehicle. Figure 6.1 shows how this might work in practice.

Capital can come from financial institutions such as banks and pension funds that *loan* money and are paid a fixed rate of interest, or from equity investors that own a share of the infrastructure and are only paid if the project makes a profit, once everyone else has first been paid. Banks and pension funds are prepared to accept a lower rate of return on their investment than equity investors but are unwilling to take on risks. Projects that carry little risk, such as building a gas pipeline that has been agreed beforehand with the regulator, can borrow money at 1 or 2 per cent more than (solvent) governments – at the time of writing goverment borrowed at 4.5 per cent. The only risk the bank faces is whether the business will go bust and default on its loans. Credit rating agencies assess this likelihood. New companies find it difficult to access cheap finance and pay more when they borrow. Unlike banks, equity investors are prepared to take on some project risks but need a higher profit margin to tempt them. A venture capitalist is a special type of investor who seeks out innovative new ideas. They look for a high return in exchange for providing guidance as well as capital.

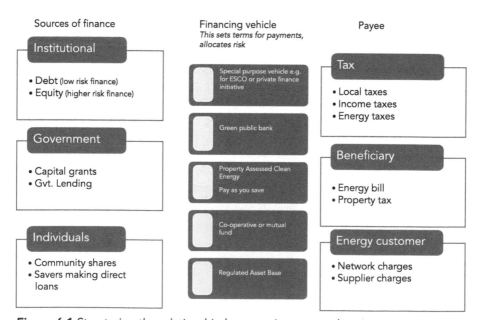

Figure 6.1 *Structuring the relationship between investor and project*

Government grants are an important source of finance for community energy projects. Government provides capital to community energy either as an outright grant or through lending money to the project at below-market rates of interest through a publicly owned bank. The grant might subsidize part or all of the cost of energy efficiency for poor or vulnerable people – in the US there are state weatherization programmes (largely funded by federal government) (see Box 6.2). In Denmark, government met part of the cost of connecting poor households to DH networks. In the UK there have been various capital grants for small domestic renewable plant such as the Low Carbon Building Programme. Below-market interest rates are common in many countries. The publicly owned German bank KfW Bankengruppe lends at substantially below commercial rates. Government guarantees KfW loans and also provides the bank with grants, and

Box 6.2 The US's weatherization programme provides grants for energy efficiency

President Obama made a promise during his election campaign 'to weatherize at least 1 million low-income homes each year for the next decade, which can reduce energy usage across the economy and help moderate energy prices for all'.

With this in mind, in 2009 the federal government assigned 'Energy Efficiency and Conservation Block Grants' (EECBGs) to cities with populations above 50,000. The funds, amounting to $6.3 billion in total, were to stimulate local efforts to reduce energy use through insulating the homes of the fuel poor. As well as trying to improve the energy efficiency of low-income homes, it was also designed to provide decent paid work and skills to poor communities.

Paul Zabriskie from Central Vermont Community Action Committee (CVCAC) found himself in the happy position that his weatherization programme budget had swelled by 40 per cent. This has allowed him to increase the number of crews of workers. But the form filling has been a headache: 'government attaches cables, rather than strings to the money'. Most of the conditionality is associated with means testing to ensure the individual is in financial need. Eligibility has to be determined by CVCAC: 'This is America, you have to pretty much strip naked to access services. Throw your pride out of the window. But we try not to make it a degrading exercise.'

As a result of the 1931 federal Act on prevailing wages, workers have to be paid the locally prevailing wage. This has been a headache as many voluntary organizations don't know what the local prevailing wage is. The result is that the recipients have been very slow to use the funds. The recurrent request from them is for examples of other local governments that have already done what they have proposed to the US Department of Energy as their action programmes by accepting the funds.[8] However, the 1931 Act does not apply to residential construction, so lack of information about the prevailing wage is clearly not the only reason for the slow roll-out of programmes. Lack of sufficient skilled staff is often a greater obstacle.

spares it corporation taxes. It's quite common for the subsidy to be paid as a tax credit, allowing companies to reduce their liability for corporation taxes or rates. A slight variation on this is where government takes on some of the project risks instead of handing out cash. For instance, government sometimes covers excess liabilities for accidents beyond a certain level – handy for nuclear power plants, which would otherwise be uninsurable. Such facilities could be useful for offshore wind too, to insure against delays caused by bad weather.

One major problem with government funding is its volatility as a result of swings in political fashion and the government's budget situation. For example, the UK Warm Front programme for the insulation of social and low-income housing saw its £345 million budget cut by 60 per cent between 2009–2010 and 2010–2011. It will shrink to zero in the following two years. Such swings in funding make it impossible for local governments and firms dependent on the grants to plan their investment programmes in concert with plans to upgrade estates or roll-out networks. Should a company with a three-year contract to deliver energy efficiency services to communities invest in innovative techniques and practices, or should it maintain a defensive posture, hiring staff on short-term contracts and using existing technologies?

Individuals can also contribute to community energy projects, either by lending money to a project or by taking out shares in a co-operative or mutual company. Such social enterprises are a mix between a business and a charity. Quite often such schemes will stipulate that the shares cannot be freely traded: perhaps locking in the investor for a period of time, or only paying interest or dividends once the project is generating sufficient revenues. Danish and Swedish wind farms and DH networks frequently raise capital in this way. These were described in Chapter 5. Here individuals invest their own savings in the community energy system. As well as providing capital on a more flexible basis, this also co-opts their support for the scheme. This means using soft political influence within the community – talking approvingly of the project with friends and colleagues identifying the venture as being borne from the community, rather than exploiting the community.

It's common for there to be more than one source of finance for any project, for instance with government providing grants for some of the cost and the rest being obtained from institutional or individual funders. It is also common to re-finance projects so the baton is handed from one source to another. While a renewable energy facility is being built, it is at its most risky – what if suppliers go bust, or there's an accident, or a vital part is delayed, or poor weather plays havoc with the timetable. Typically the project developer funds it from his balance sheet and from short-term (and expensive) bank loans. Once the plant is complete and has been switched on, the risk profile might be reduced sufficiently for a risk-averse pension fund to take it on.

Different groups can have ultimate responsibility for paying back the costs of the investment. Figure 6.1 shows it could be government, but it would be unusual for government to repay the costs of community energy loans through taxes, since government can borrow more cheaply than the private sector so it's more efficient for government to raise the capital itself. Of course, there are always exceptions to every rule ... sometimes for no good reason. The UK government made extensive use of the PFI to keep public loans off the public balance sheet. Instead of issuing bonds itself, the government asked banks to borrow money for a project and it simply paid the interest. This ploy ultimately failed when the national statistical office refused to ignore the existence of these 'loans' and put them on the government's balance sheet, to the embarrassment to the government. Such PFI deals have financed the construction of energy-from-waste plants that generate electricity (and could be used to produce heat). Another example of taxes, this time energy taxes, used to repay loans is the widespread use of feed-in tariffs (FiTs). The German FiT[9] raises money through energy bills, which is used to compensate investors in renewable electricity and CHP. As with other FiT systems around the world, a complex system of equalization payments takes place behind the scene. These arrangements ensure that an energy company with a high proportion of customers in receipt of feed-in payments are not put at a commercial disadvantage.

Taxation, as well as being an expensive way of raising finance, also raises distributional issues: if taxpayers are responsible for repaying the project cost, the question arises, which taxpayers? Repayment using nationally collected taxes forces all taxpayers in the country to pay for the infrastructure even though the benefits might only be going to a small community: some of the payees might be much poorer than the people benefiting from the investment.

There are a number of mechanisms for the communities directly benefiting from the investment to pay back the cost of the investment. If the community improvement is to the fabric or their home (for instance where a block or street has an area-based insulation programme, or community energy infrastructure) the loan might be paid back either through a charge on their energy bill (often called on-bill financing) or through an additional property tax. This is an attractive mechanism in that it is fair: the person who is benefiting from the improvement is ultimately paying for it. It also makes more intelligent use of the finance: providing (hopefully) cheap capital to someone who might otherwise not have access to capital, potentially allowing an improvement at no upfront cost to the beneficiary.

Energy customers can fund cost-effective projects in another customer's home. This is a common way of funding low-cost energy efficiency in the US and UK – this might include projects such as subsidized light bulbs, cavity wall insulation and loft insulation. We have drawn a (slightly artificial) distinction between this and the FiT 'tax' described above. Because many energy efficiency measures are low cost, it is feasible that most customers will receive some benefit from the programme, though there is some redistribution between customers. The low-cost measures are cost-effective and so the programme should result in lower energy bills across the entire customer base.

We argue throughout this book that local communities should decide what community energy infrastructure they need. But the investors will also want a say in how their money is spent. Investor-owned energy utilities and banks have a duty to achieve the highest rate of return on their investments. The investment they make in renewables and energy efficiency is because of price incentives or regulations to stimulate such investments. This contrasts with government and individual investors. Individuals and governments are not averse to seeing a good return on their investments, but this isn't their only criterion.

We would like to see much more local control of energy investment, but does that mean the local community has to foot the bill? The plausibility of relying on local funds varies across countries. In the US, most taxes are raised locally and the federal government provides only a small fraction of the funds needed for operations of the 50 states, 2000 counties and tens of thousands of local government bodies. As a result, there is a well-developed system of sub-national tax and fee collection. In the UK, local government is allowed to borrow money from national government as long as the revenues from the investment can pay the interest on the debt. Money is lent to the local government from the Debt Management Office, which is a branch of the national government. Local government is audited to ensure that borrowing is undertaken responsibly. Some local governments have also used the European Investment Bank to fund large energy infrastructure projects. In the US, by contrast, there is no nationally established limit to local borrowing, and there are even federal programmes to subsidize and encourage such local financing independent of individual state policies. While the US states, in principle, have the power to limit such borrowing, permission to borrow is almost never denied in practice.

The Financing Vehicle – Managing Risks

Uncertainty kills investment prospects. Novel investments in community energy have a myriad of risks and uncertainties that could blow up in the investor's face. Few public officials, elected or appointed, want to take a gamble on an investment that might become an expensive fiasco. The smaller and lower-level the governmental unit, moreover, the more likely it is that the public officials will look up for guidance. In the US, where full government powers are given to really small communities, that nominal power is often a source of stasis and resistance to change. The expertise to consider options may not be available. The prospect of change is thus scarier than it would be in settings where the alternative paths forward can be carefully compared.

This point is even more true of investors than government officials, even investors whose role is nominally to take on exceptional risks. The joke about venture capitalists – those who sink funds into innovative high-risk, high-return start-ups and small firms that have yet to show profits – is that they like to be first in the queue for the second-of-a-kind project. They want to build on others' experiences in order to reduce their uncertainty. They are willing to take on risks, so long as their magnitude is known.

The choice of appropriate financial vehicle is really a means of sorting through the morass of things that could go wrong and allocating these risks (and also the potential for rewards) between the investor and community as cheaply and as straightforwardly as possible. It involves selecting, or maybe combining, different sources of capital or absorbing the risk that otherwise drives up the cost of capital. Who bears what risk? How are benefits shared? What is the precedence among investors in getting paid off, especially in the event of the failure of an initiative?

In the idealized case, communities would choose a strategy by:

- planning the infrastructure and capital needed (and political capital, in some instances) and integrating these plans with broader ambitions for the community, for instance, creating jobs and local economic development;
- determining the returns on investment available under best-possible future conditions; this will include non-energy benefits and ensure the vision is shared so the programme is widely owned;
- determining the alternative futures that might exist – the conditions generating the risks – this might include project over-runs, overly ambitious take-up rates, changes in prices and poor performance;
- managing the risks – perhaps by using financial markets, co-opting opposition or handing over the risk to someone else (for example contractor, insurance company);
- sourcing the capital and investment vehicle to reduce the cost and risk of the project.

There are a variety of energy service contract models. In some the ESCO takes over the risks of an energy investment failing to perform.[10] Here the ESCO is paid from the savings they make, rather than the value of kit they sell. Such ESCOs assess buildings' energy-saving potentials and then undertake the necessary retrofits, from whole heating and cooling plants to insulation and control systems. The ESCOs needs to guarantee energy cost savings that will cover the debt service costs of the loans taken out to finance their operations – the firms have minimum scale projects they will undertake and thus are not available to individual home-owners.[11] Box 6.3 shows how these contracts have been applied in Upper Austria and Berlin. In both cases the ESCO, rather than the building owner, provides the upfront finance. In Berlin, KfW has lent money at low interest rates to the ESCO for this purpose.

ESCOs are most effective if they operate at scale – the Berlin model recognizes this by requiring buildings to be aggregated. Any construction activity has substantial fixed costs for procurement, management and coordination, for bringing equipment to a site, and the like. As those costs are spread across larger projects, the unit cost (per building, dwelling unit or square metre) will be reduced.

However, neither of these ESCOs have undertaken particularly risky investments. There is a limit to the type of risks the ESCO can bear. The model can't easily accommodate occupiers that fritter away the efficiency by cranking up the heat. The energy market in which they are embedded restricts their freedom of manoeuvre. ESCOs have to buy and sell power through the power markets but don't have the ability to insure themselves against fluctuations in the wholesale price as easily as a large integrated generator supplier. They have to buy gas at wholesale gas prices, but without the economies of scale that a major supplier enjoys. There are clear trends that explain why the ESCO business model does not include buying insurance coverage for the risk that energy cost savings will not cover debt service charges. Under normal circumstances, one would expect

Box 6.3 Energy service contracts in Berlin (Berliner Energieagentur) and Upper Austria

In 1992, Berlin set up its own Energy Agency. It is described as an 'Energy Services Undertaking' and is a partnership between the Land (federal state) of Berlin, KfW and two private companies Vattenfall and GASAG. The Agency annually lends €2.5 million to business and community-level projects.

One of its activities is to help commercial offices and the public sector reduce the energy consumption of their buildings. The Agency acts as a project manager, helping to devise, negotiate and oversee the process. One of its programmes agglomerates the energy used by nearby buildings so the combined value of energy used is at least €200,000. Each pool comprises between 4 and 400 buildings. Between 1996 and 2008 it had arranged 24 such pools covering 1300 buildings. These contracts are then tendered on an energy service contract basis; firms such as Honeywell and Siemens might bid. The ESCO pays for the retrofit up front and building owners pay them back over an agreed period – usually 8 to 12 years in annual instalments from the energy savings. Typically around 97 per cent of the annual savings are paid to the ESCO in the early years.

Once the contract is completed, the building owner realizes the energy saving. Altogether the ESCO has invested €49 million into the pools and reduced their energy consumption by €11.3 million or 26 per cent. The ESCO usually invests in insulation, CHP, lighting and better controls.

Upper Austria operates a similar programme. The Energy Contracting Programme offers financial support for energy efficiency, with work carried out by an ESCO. Building owners are spared any upfront investment. The ESCO guarantees that it will reduce energy costs by a certain percentage every year. It is in charge of financing, installing and, where necessary, operating and maintaining, energy-saving measures and systems. The money saved by lower energy use is initially shared between the ESCO and the building owner. Once the cost of installation is recouped – which takes 10 to 15 years – the saving goes entirely to the building owner (unless there is an ongoing cost of operation and maintenance).

Upper Austria was the first region to use a third-party financing approach for energy efficiency. At first, the Energy Contracting Programme was restricted to public buildings, and to energy efficiency rather than renewables. It was expanded to businesses and to renewables in 2002. The programme offers financial support up to 6 per cent of the energy investment, or up to 13.5 per cent of the investment costs for renewable heat (up to a maximum of €100,000).

that businesses would insure themselves to protect against too many claims on their guarantee. But the ESCOs' assurance is absurdly conservative: they will only propose and install those energy efficiency measures that will permit savings to cover costs on the assumption of constant energy costs. Since average energy costs per year have risen steadily over recent decades, what this business model actually means is that the energy efficiency investments ECSOs offer are less extensive than would be warranted by a break-even analysis that incorporated price trends.

ESCOs can potentially work on homes as well as on commercial properties – for example at the community level across housing estates. The ESCO can invest in enhancing the energy efficiency of homes, perhaps by installing high-cost insulation measures or CHP systems. Bundling a mixture of housing and commercial buildings allows a more even demand for heat and power over the course of a day, resulting in better utilization of heat infrastructure. However, such community ESCOs are not common. One practical issue is that unlike office workers, homeowners are used to paying for domestic energy according to their metered energy use. If their insulation is improved, they might react by keeping their homes at higher temperatures. From a social point of view this might be desirable but from the point of view of an ESCO, which depends on the reduction in energy consumption to reward its investment, it is a major problem. Community ESCOs can also be fiendishly hard to negotiate. Imagine trying to bring together a group of people – some might be renting their homes on a short-term basis, some are owner occupiers, some are well-off and can afford to invest money themselves, some are on benefits and in receipt of low-price energy, some might hate the idea of change, and some might have a deep green agenda. Bringing all these people together takes time, and the discussion might be inconclusive. The agreement has to be between the ESCO and whosoever resides in the home and has to endure even if people move home. So someone has to make a decision and this might not be popular with everyone. The UK established a £50m Community Energy Programme for three years but most of this money went unspent because it proved impossible to organize and spend the money within the timeframe imposed by the finance ministry.

One project that did obtain funding was a scheme in Aberdeen (described in Chapter 5) to install CHP in a tower block. Half the capital cost of £1.6 million was from a government grant, the rest was bank borrowing and avoiding expenditures on individual heating systems in each flat. The local authority established an arm's-length not-for-profit ESCO owned by the council, which bought in private sector expertise and services to perform specific tasks. The ESCO charged residents a flat rate for their heat, so the residents did not bear any energy price risks. This was motivated by a slightly paternalistic desire to stop residents worrying about keeping warm. A community ESCO scheme in Woking established a public–private joint venture Thameswey owned by a Danish partner Xergi Ltd (81 per cent stake) and Woking Borough Council (19 per cent stake). The set-up costs for the company were substantial but the venture has undertaken some immensely innovative schemes.

A Green Investment Bank

The UK's coalition government has promised a green investment bank. The Treasury is not persuaded that a new government institution is needed, believing that normal banks will lend viable projects money. Many environmentalists see the bank as a totemic indicator of whether or not the country is serious about environmental investment. Both parties cannot be correct.

Box 6.4 KfW banking group

KfW is the German development bank jointly owned by the federal government (80 per cent) and the states of Germany (20 per cent). It was set up in 1948 to finance Germany's reconstruction following the Second World War. It operates independently of the government but is required to encourage economic, social and ecological development worldwide. In 2009 KfW borrowed €74 billion from international capital markets through issuing bonds. It is regarded by ratings agencies as one of the safest institutions to lend to in the world, and so can borrow cheaply, and also borrow many times more than the capital the two shareholders have paid into the bank: the bank currently borrows and lends 38 times its paid-up capital.[12] The bank pays no dividends to shareholders and also is exempt from corporation taxes, allowing it to lend money at below commercial rates of interest. It is forbidden from competing with commercial banks.

It is now active in financing renewables and low-carbon enterprises and around half its loans are for housing and environmental purposes. It is the world's biggest lender to renewable energy projects. It has offered low-interest loans for refurbishment since 1990. From 2001 loans have been available specifically for energy efficiency improvements. In 2006, the KfW programme on energy efficiency was almost tripled to around €1 billion a year (and subsequently increased to €1.5 billion). Since 2007 direct grants have been offered alongside the loans. Such is its dominance in energy efficiency lending that it now sets the standard for low-energy retrofit in Germany – the KfW-40 and KfW-60 have become standards for energy efficiency. In Germany, the KfW development bank raises AAA-rated, government-backed bonds for its energy efficiency household loan programme. It then subsidizes these loans so they can be offered to the consumer at a rate of 2.65 per cent,[13] much lower than the market rate of interest, or indeed the rate the government can borrow at. This is supported by further grants and some regulation. As a result, the programme is achieving 100,000 retrofits of residential homes a year.

KfW's CO_2 Building Rehabilitation Programme is carried out by local banks and offers loans to owner occupiers, landlords, housing companies, housing co-operatives and local government, for energy efficiency, CHP and renewables. Some states support the programme with reduced interest rates. In 2007, grants for owners of one- and two-family houses were also made available. There have been specific KfW loans and grants for non-profit organizations, local authorities and associations of local authorities available since January 2007.

We believe that a green investment bank tasked with funding community energy is essential if large-scale community investment projects are ever to get off the ground. ESCOs and local government have to go through immense and unnecessary hurdles to borrow money for bankable projects – the Aberdeen ESCO ended up borrowing from the Bank of Japan as UK institutions were unwilling to lend. A green investment bank must swiftly develop expertise in low-carbon projects, borrow money from the money markets and crucially be guaranteed by government, to access low-cost finance from pension funds, as soon as it starts operations. Existing financial institutions such as pension funds and banks have neither the experience nor the appetite to invest in unconventional investment projects with little track record. This hesitation puts a self-fulfilling blockade on environmental progress. There is already significant experience of public-supported lending to environmental investment. Most famous – among Germans and environmentalists at least – is the German KfW explained in Box 6.4. The German government guarantees lending by the bank. The bank subsidizes interest rates allowing some lending to occur at 2.5 per cent, which means that even long-lived, slow-burn projects like DH networks start to make commercial sense. A number of the international development banks are already active investors in community energy. The European Bank for Reconstruction and Development set up in 1991 to invest in the former Eastern Bloc has a substantial array of funds for low-carbon investment in Eastern Europe and Central Asia. Often money is lent to small retail banks within the host country, to develop in-country skill and expertise in assessing such projects. The EU's European Investment Bank is also a major lender. Neither of these banks is guaranteed by government but both have a long track record of participating in the money markets and can borrow at rates similar to government. In the US, the Rural Utilities Service lends money to rural electrification programmes at federal interest rates.

Upfront Payments – Dealing with Uncertainty

Let us eat and drink; for tomorrow we shall die.
The Bible (Eccl. 8:15 and Isaiah 22:13)

People hate spending money now for benefits they will receive a long time in the future. This is even true of economists who should know better. Prashant's desire for instant gratification has made taking a lunch box into work impossible – since it is consumed shortly before noon. Economists dignify this short-sightedness by saying individuals have a high rate of discount, as though that makes it OK. When they're being super-smart, they say people have hyperbolic discount rates, meaning they care a lot about now but are largely unconcerned about the difference between ten years away and eleven years – both are simply points in the distant future.

This makes getting people to invest in pensions, or energy efficiency for that matter, difficult. This is where schemes that avoid the need for any upfront payment come in. Box 6.5 sets out two schemes: PACE from US and the Green Deal from the UK, that allow homeowners to access capital to invest in energy efficiency or micro-renewable energy without any upfront payment. Instead they pay an agreed regular charge either on their local property tax (PACE) or as an extra fixed charge on their energy bill (Green Deal). The Green Deal also promises that the savings from the energy efficiency are greater than the monthly charge being levied by the company installing the energy efficiency, which means that the homeowner not only avoids paying upfront charges, to flatter his impatience, but he is also no worse off from putting in the measure, appealing to his prudence. We are actually quite sceptical that there are many investments that can be made in the average home that are capable of achieving the latter condition and that this will quietly be dropped in a year or two, once political face has been saved and ministers have moved on. Crucially, the agreement to pay the charge on energy or properties bills binds not just the homeowner but future homeowners that take over the house.

Box 6.5 Property Assessed Clean Energy at Palm Springs, California and the Green Deal, UK

The initial capital cost is a barrier to investing in renewable energy and comprehensive energy efficiency improvements to homes. PACE financing is a financial model that addresses the upfront cost issue by attaching the loan to the property rather than the home's owner. This way the homeowner is confident that if they move home before the loan is paid off they're not stuck in the situation where the new homeowner benefits from the investment without paying the cost.

The PACE mechanism was devised by Berkeley City specifically to provide residents with the means to finance the installation of solar photovoltaic panels at no upfront cost. The average requested loan when the scheme was introduced was $28,000 but this could ultimately be paid off from the generous subsidies offered in California for electricity produced from photovoltaics. The PACE model has been widely copied and the largest scheme is operating in Palm Springs. In Palm Springs the programme lends residents money for 20 years at 7 per cent interest. Between August 2008 and early 2010 the city authorities approved 220 applications; they have issued bonds of $1.1 million and $2.5 million to raise the finance. The interest is paid by the homeowner through a lien (surcharge) on the property tax levied by the city authorities. The new owner inherits the lien if the property is sold before the loan is paid off.

While the PACE scheme has been widely copied by many states, it has not been hugely successful. For one thing the rate of interest charged is not particularly attractive. Most of the people who applied for the original Berkeley scheme did not ultimately sign up for the programme since they could borrow money more cheaply by simply extending their property loan. The problem has been that small cities have simply not had a large enough number of loans to pool. This means the cost of the set-up and administration of the debt is large compared to the value the authorities are borrowing. The mortgage lenders have been deeply unhappy with the scheme since the lien is regarded as senior to their mortgage. This means that if the homeowner throws in his keys, the city will get paid ahead of the mortgage company. This point has been a bit of a showstopper. The two federal agencies that act as backstop guarantors for home loans in the US, Fannie Mae and Freddie Mac, have protested about the greater level of risk they are exposed to. (Given they are already insolvent and have been under the conservatorship of the federal government since 2008, one does wonder what there is to fear from a little more risk.)

In the UK the coalition government is launching the Green Deal. This flagship policy means that homeowners can install energy efficiency measures in their homes at no upfront costs. Instead a private company, local authority or social housing provider borrows money from the money markets, installs the measures and is paid back from the savings from the energy efficiency measures. The homeowner repays the provider through a charge on the energy bill. At the time of writing, government ministers are promising that the homeowner will be better off from the investment, so only measures that have rates of return higher than the cost of finance will take place. The government is also promising that the Green Deal provider cannot take the creditworthiness of the individual into account when making the loan. This means that the Green Deal provider will be paid once the measure has been correctly installed. This is to reduce the riskiness to lenders of, say, a new owner deciding he didn't approve of the previous owner's decision to install some grotesque-looking external insulation and unilaterally deciding to cease paying. A recalcitrant customer would be cut off by their utility.

At the time of writing it was unclear whether the problems encountered by PACE – lack of volume of loans driving up interest rates and the baulking by other lenders of the energy efficiency charge's favoured position – will unseat the programme.

Work on piloting the Green Deal (then confusingly and optimistically called the Pay As You Save model) showed that very few measures in the home would actually 'pay as you save' if the installer had to pay commercial interest rates, but thankfully this didn't much matter to consumers. Most people didn't know exactly what they spent on energy, and so long as the monthly repayment sounded reasonable they weren't that concerned whether it saved more than it cost. They were motivated by a desire to improve the environment and the comfort and value of their home. What consumers actually prized the most was being spared the hassle of researching technologies and finding reliable suppliers.[14]

One significant risk that faces investment in community energy is that beneficiaries procrastinate in signing up to a service like DH. A community might agree the network is built, but individuals will be slow to sign up to the service when it is complete. This ruins the economics of the scheme and exposes everyone else to paying the fixed costs of the scheme. This situation will be all too familiar to anyone that has ever tried to organize an office party. Everyone puts their name down to attend and the price is negotiated with the venue on that basis, but as the day approaches people start to cancel, unseating the viability of the event.

Society has created mechanisms such as deposits or advance payments to ensure risks do not fall on others.

Similarly, the idea of attaching repayments to the energy bill (or property tax) does not have to be restricted to investments that homeowners voluntarily enter into. It could, in theory, be applied at the community level as a means of compulsorily recovering the cost of community infrastructure like DH networks, local generation plants or improvements to a tower block to allocate the cost to all beneficiaries and more controversially *potential beneficiaries*. When community infrastructure is financed in this way, it is functionally equivalent to the way utilities presently finance enhancements to gas or electricity networks. These expenditures are authorized by the energy regulator and the costs recovered through a return on the RAB. This finance mechanism could also be used to cheaply borrow money from wholesale money markets or a green bank. Prashant and Ed Mayo argue this approach elsewhere.[15]

Co-operative ownership is also a good means of reducing the risk of backlash from local communities since the communities themselves have pooled their interests and taken over joint ownership of the community energy infrastructure. We talked in Chapter 5 about the Baywind co-operative in the UK that has been responsible for developing community-level wind projects, and the much larger Danish wind and DH co-operatives. In the former, the capital is largely supplied by well-off individuals, while in the latter, funding either comes from individuals or from government-guaranteed loans. Co-ops can galvanize local community support for some types of projects. Such forms of local participation work up to a certain scale of project – where all affected parties can gather and discuss and participation is genuinely inclusive. It is more difficult for large projects, say to develop energy plant for a whole town, or entire systems of interconnected works where the decisions and trade-offs are more technical. Also mutual or co-operative organizations – where members volunteer to work together – don't usually have the powers to compel people (aside from contractual obligations) or inspect and enforce its decisions. In these circumstances we have to turn to investment by local public authorities.

Investment and Local Government

Local government is the big beast in any community and inevitably needs to be a major player in any local energy infrastructure investment programme. It has substantial resources to invest in community infrastructure, powers to raise taxes, it is a major owner of commercial buildings and also builds and rents out social housing. But working with local politicians presents its own challenges, one of which is that local energy production is usually not uppermost on their list of concerns.

On 2 September 2008 Toronto mayor David Miller made a speech on the Tower Renewal project. The City has 1000 residential tower blocks built before 1984. Retrofitting these blocks could reduce electricity and gas consumption by half (relieving the problems the city is having with its transmission lines), saving around 700,000 tonnes of CO_2 per year. This is the same as is emitted by a medium-sized town. Each building will cost several million to retrofit. In the speech he articulates his motivation for the programme:

> Why are we doing this? Toronto is a city of towers ... tower renewal seeks to reduce greenhouse gas emissions by cladding buildings to reduce emissions by 50 per cent, it's about creating local green jobs, it's about upgrading green space around the buildings and installing green roofs to help biodiversity and water management, and its about taking greater responsibility through on site management of waste. We want to create expectation of green sustainable communities.[16]

The process of agreeing this programme involved consultation with dozens of local organizations, and not surprisingly it had a multiplicity of objectives. It is not just about energy management – and even if it was he probably would never have admitted the fact. Local politicians are voted in by one section of their community, but they have to govern on behalf of their entire community. The speech was made outside, near one of the towers; his audience included people living in the flats who wished for warmer, cheaper-to-run flats, construction professionals who he wanted to inspire into producing excellent work, and his electorate who he wished to woo through promises of jobs and improved inner-city environments. At the time Miller chaired the C40 group of mayors concerned about climate change issues.

How would this ambitious scheme be financed? The cost of the programme might be $2 billion over the next 10 to 15 years. Most cities – Toronto is no exception – do not have this kind of money lying around. And anyway, since the blocks were privately owned, why should the city pay? Consultants recommended the city create a special purpose organization (The Tower Renewal Corporation) whose job it was to raise finance from the capital markets to finance each block renewal (costing some $2 million each) and manage the renewal itself.[17] The building owners would pay back the Corporation over the next decade or two. This approach aggregates the individual projects to make them large enough to take to the money markets. But banks and pension companies are unused to this type of project and would still charge a risk premium. To reduce this, Toronto would put in a modest capital injection (between 3 and 10 per cent of the fund value) to reduce the riskiness of the bond and reduce the interest rate passed on to the property owners. This they clunkily call a credit-enhanced capital pool. This works for the landlord as long as the value of the tower is increased through the renewal project and if residents are prepared to pay a premium for the better facilities, living conditions and lower energy bills. The report[18] recommended that to manage the risk of default by the property owner the loan be collected back through property taxes, by putting a lien on the property tax. This is the PACE mechanism described in Box 6.5.

Box 6.6 The Rotterdam Climate Initiative

Rotterdam has a concentration of heavy industry, oil refineries and power stations and 600,000 inhabitants. It is a major source of greenhouse gas emissions. The Netherlands is vulnerable to sea-level rise and Dutch politicians and businesses have a shared concern about climate change. Rotterdam is an active member of the C40 cities climate leadership group.

The City established the Rotterdam Climate Initiative (RCI) in 2006, in partnership with the port, a not-for-profit called Deltalinks and the Dutch environmental protection agency. RCI was the initiative of Ivo Opstelten, mayor of Rotterdam between 1999 and 2009. His motivation was partly to win greater public support for the expansion of the port and partly to 'rebrand' Rotterdam in order to attract more investment.

RCI is well connected – it works with the Clinton Climate Initiative, and its board is chaired by former Dutch Prime Minister Ruud Lubbers. It has a target to halve the city's emissions by 2025. To achieve this, it works extensively on energy efficiency (retrofitting of existing stock), renewables (predominantly biomass and wind energy) and DH.

Stephen visited Rotterdam in July 2010. RCI initially received €50 million from the city council, which had been raised by selling waste incinerators. This was not enough to fund RCI's ambitious plans, so it has always had to cultivate partnerships.

RCI is focused on the existing building stock. Housing corporations own half of the homes in Rotterdam and the city council owns 4000 homes. Many are apartments, making it difficult for residents to get a decision allowing them to improve the building fabric. RCI has persuaded housing corporations to enter agreements to retrofit existing buildings, and half have now done so. The average efficiency of buildings is improving by 2.5 per cent a year. Woonbron, the most active housing corporation, allows people to choose whether to buy or rent their property. It has established an NGO to encourage its residents to improve energy efficiency. This approach is supported by RCI. In its judgement, NGOs are the best organizational format to promote energy efficiency – energy companies do not want to be involved as the payback times are too long.

DH is not yet widespread in Rotterdam but new biomass DH plants are being planned. There are helpful national policies on biomass in The Netherlands. In 2009 the Dutch government adopted criteria to define the sustainable use of biomass. Biomass plants receive no public subsidy unless the heat is used efficiently. Rotterdam has a local law that any building constructed has to be connected to a DH system if one exists. It took five years to get this adopted as a law, as most people assumed that natural gas would remain plentiful and cheap.

The City is planning to extend DH through an infrastructure company, owned by the city, and an operating company, owned by the City and E.ON Benelux. Heat will be supplied by a waste incinerator. The city council is providing €38 million of capital and providing surety for the loan of €149.5 million.

Not only does each investment project have to embody a multiplicity of different objectives – there are also a lot of other matters in the in-tray. Other speeches Miller made in that three-month period included comments on Remembrance Day honouring the casualties Canada suffered in the Second World War, a response to Mumbai terror attacks and comments about the $280 million contribution the province of Ontario had made to the city's infrastructure.

Local politicians have to articulate the case for community energy in a way that will be politically sellable in their locality. In some cases, the commitment may be to addressing climate change, but in others, the drivers may be purely economic gain. That gain may not even relate directly to energy costs.

Rotterdam's efforts to reduce its GHG emissions by 50 per cent, for example, appear driven as much by business development considerations as by climate change concerns (see Box 6.6). The city and its region are pursuing a reputational change for what is now Europe's largest port. The economic development potential of becoming a green power hub (it is promoting itself as a potential CCS centre) and of attracting other businesses by offering a green, not gritty industrial, image would make energy efficiency efforts economically attractive, even if power cost savings did not pay for the improvements in buildings (which they do).

This process of inserting community energy into the job description of the local politics can produce profound and long-term changes. As shown in Box 6.7, Upper Austria has invested substantially in developing heat networks that are fed using locally produced forest products. This strategy has created 10,000 jobs, many in rural villages with few alternative employment opportunities, and enhanced the region's energy security. The programme has managed to redirect EU and national subsidies for agriculture and capital grants for heat networks to finance an integrated package of investment in biomass-fired CHP. Regulation has also been used to compel the public sector and commercial sector to invest in renewable heat sources such as solar thermal.

Local government has access to unconventional sources of finance that aren't available to communities. Local government has tax-raising powers, which vary tremendously between different countries. These are very circumscribed in the UK, limited to residential property tax that has to be applied uniformly to all citizens. Local tax raising is more varied across US cities and counties. Local governments also have the powers to charge for council services.

Box 6.8 outlines how the City of Babylon, in New York state, employed the scientifically illiterate but legally kosher redefinition of CO_2 as solid pollutant to raid one of its funds. Toronto has developed the unique deep lake cooling system, which draws water from the bottom of the lake Ontario, which is a constant 4°C throughout the year, to cool a hundred commercial buildings in the downtown area. The project was financed using $250 million drawn from the City's pension fund, which has a controlling interest in the scheme. Employees of the City now have a rather direct financial interest in ensuring that the cooling system remains in good financial shape. In the UK, a number of councils have used reserves to finance community energy efficiency through lending funds to residents via 're-volving funds'. Kirklees Council in Yorkshire pioneered this approach, recycling the windfall gains from the reduction in employer taxation that accompanied the introduction of the UK Climate Change Levy.

Box 6.7 Financing of heat networks in Upper Austria[19]

Austria's regions play a significant role in energy, particularly regarding building efficiency and heating, including DH. Upper Austria covers 12,000km^2 and has 1.4 million inhabitants. It is highly industrialized but there is a signif-icant farming community too. Most Austrian farmers own both farmland and forests, but in mountainous areas agriculture is not particularly profitable. Energy from biomass has been strongly promoted across Austria. The prime objective has been to support agriculture. Around 10,000 jobs are supported by the renewables industry in the region.

Since the early 1990s, the regional government has been actively and effectively promoting energy efficiency and renewable energy, particularly renewable heat: biomass, solar thermal and heat pumps. Around a third of energy used in the region is renewable: around 78 per cent of electricity and 45 per cent of heating. Hydropower and biomass are the chief sources of power and heat respectively (though solar hot water and heat pumps are also significant). Modern biomass boilers that comply with air quality regulations are expensive. In order to make efficient use of this biomass it is combusted in CHP plant with DH to share the heat, instead of individual boilers. There are around 300 DH networks in the region, which link 8000 buildings capable of supplying 225MW. Most of the networks are owned by co-operatives of forest owners that also supply the fuel. They were built over the past 20 years.

The region has a renewable heating law, which mandates that all new and refurbished public buildings (since 1999) and all new private sector buildings larger than 1000m^2 (since 2008) to install renewable heat measures. Programmes to extend DH using biomass are managed by regional goverment, but the national ministry of agriculture provides around half of the money. Subsidies of up to 30 per cent of the eligible costs of biomass installations are available from the federal government (with regional governments offering up to a third of the cost as additional subsidy). In 1988, the ministry spent €950,000 on biomass DH; in 1993 this had risen to €7.3 million. In 1999, €11 million was provided by the ministry, €7.3 million from the Länder and €5.1 million from EU regional funds.

Farmers have also been offered money – subsidies and low-interest loans – to encourage them to install biomass facilities and connect them to DH (in addition to the money they get through the EU Common Agricultural Policy for growing the biomass). Farmers or farmers' co-operatives have been able to get a higher percentage of installation costs than have private companies. This has led to some energy companies seeking to enter the heat market to set up co-operatives with farmers. The federal government has also given grants of €800 to householders for biomass heating. In the 1990s, it spent around €5 million a year helping small firms innovate in their use of biomass.

Box 6.8 Creative use of local resources: The Babylon project

Few local authorities find themselves in the situation of sitting on a pile of cash. The problem arises when it's in a box labelled: 'Solid waste fund. Not to be used for any other purpose'. Babylon, NY, US, found itself in this situation in 2006. Political leaders realized they'd been accepting waste fees way in excess of what could usefully be spent preventing solid waste. Nor would the authority create much political goodwill by returning the money to citizens.

They decided to tap the fund for loans to private homeowners for energy retrofits. To do this the City had to redefine carbon as a solid waste, returning the city to a pre-Enlightenment level of scientific ignorance. The redefinition of CO_2 as a solid waste allowed Babylon to assist homeowners retrofit their homes. Homeowners faced two barriers to investing in energy efficiency. First, with the average homeowner moving every seven years, such an investment would have to be paid off in three to four years and/or would have to increase the resale value of homes to be worthwhile. Second, most would need access to capital, and that could be an issue for middle-income homeowners with already high debt burdens.

Terms for provision of the retrofit funds included bank rate interest. Repayment was assured through imposition of liens on the properties being improved, collectible in the same manner as any outstanding taxes due on real estate. Such 'benefit assessments' have a century-long history in the US: localities borrowed funds, and placed supplemental liens on properties to service their debt, to get the capital needed for infrastructure improvements such as new sewer lines, sidewalks or repaved streets. Under the programme, homeowners pay $250 up front for an energy audit that determines the work needed on a home. If the homeowner then agrees to proceed with the municipally financed retrofit, the $250 can be reimbursed out of the benefit improvement monies. The energy audit establishes the funds needed but the work is assigned by Babylon itself, so it can limit work to local contractors and provide incentives for those contractors to utilize local unemployed people, many of whom need training in the relevant building trades skills.

With 309 homes audited and/or completed the programme showed:

- The average cost of improvements to meet Energy Star-type standards was $8080.
- The average annual saving to homeowners on heating and cooling bills was $990.
- The resulting average payback period was 8.2 years.
- The expected savings to investment ratio (SIR) over the lifetime of the improvements was 2.1.

These returns were calculated under the assumption of constant costs of heating oil and electricity and thus understate expected returns that adjust for current upward price trends. Moreover, the pilot has provided Babylon with information on how to deal with contractors doing energy retrofits, standards to apply for work to be conducted, auditing needs for the work completed and the willingness of local contractors to train and employ local unemployed youth as they expand their workforces in response to the offer of new construction work from the municipality.

Local Government Using Public Procurement

Local governments are major landowners within a community. They own and manage school buildings, leisure facilities, civic buildings, halls and commercial offices. They might also own social housing. By using their purchasing powers wisely, they can create and shape the local energy and energy efficiency market. By committing to buy heat or power from a new renewable scheme – and thus provide an anchor stream of revenue for the scheme – they can reduce the project's uncertainty and allow it to access lower-cost capital. This creates a virtuous circle by making the energy from the scheme less expensive to those who might repower communities. In Upper Austria, new public buildings had to make use of renewable energy in 1999, a full eight years before the mandate was extended to new private sector buildings.

Local government can also use their permitting and local planning powers to make new homes and businesses connect up to community energy facilities. The Danish Heat Law explained in Chapter 3 is the most obvious example of this. The Heat Law mandates that homes and offices must connect up to the heat network within ten years of the network being built. This makes planned heat network schemes bankable, since the co-op or municipal authority could go to the money market and show a predictable return.

Local governments, when they act on their own and finance themselves, can accomplish far more than commercial ESCOs. In the US, local governments and independent school districts are expected to borrow for their investments. Entering into such debts is only rarely politically controversial. The Kentucky schools case, however (see Box 6.9), relies on minimizing overall costs. The requisite planning can take into consideration the trend lines in energy costs, allowing them to act 'imprudently' by normal banking standards, looking after their citizens' long-term interests. This approach to making its own mind up on imponderables such as future energy security means the education authority has integrated energy efficiency into its day-to-day business of building, and schools can innovate. Some schools are approaching net zero carbon energy efficiency operations through carefully planned investments in retrofitting older schools, building new ones, and educating staff and site visitors, including students, about energy-efficient building operations. In the process, their borrowing is on balance saving taxpayers significant sums.

The extensive reliance on DH systems in Scandinavian communities demonstrates that their know-how and political determination to deploy heat networks has allowed them to take advantage of efficiencies of scale in providing a needed resource to compact human settlements. This requires creating financial models to provide cheap capital, enduring revenue streams and institutional forms that assure customers that monopoly heat providers will not exploit them.

Box 6.9 Energy progressive?... Kentucky in the US?

It is a Southern state! It is the third largest coal mining state in the country! It generates 93 per cent of its electricity from coal! Its households consume 125 per cent of national average residential use of electricity! But in 2009, the US Energy Star programme recognized the Kenton County School District for achieving a 10 per cent reduction in energy consumption. In September 2010, Warren County, KY, opened the first zero net energy school in the entire US.

These accomplishments are not a new development and have nothing to do with concerns for emissions or global warming. The Kentucky Department of Education has been actively promoting school building energy efficiency. Schools in the state have been installing ground source heat pumps (GSHP) since 1990. Political progressivism can't explain the pattern, nor can environmentalism. One Kentucky legislator introduced a bill in 2009 that would have made it illegal to spend any state funds on energy conservation! It didn't pass.

Why, then, have Kentucky schools been in the forefront in pursuing energy efficiency?

The simple answer is a four-letter word: cost. A further explanation lies in the process of accessing capital for school retrofitting and construction. Kentucky is very decentralized, with 174 school districts educating some 640,000 students. Each district is responsible for its own finance, and capital costs are generally financed through the issuance of 30-year bonds. Such small districts would face high underwriting and interest costs for floating their own bonds, so the state's Department of Education centralizes the process of raising capital for them.[20]

The review and approval process assures districts of a common bond rating and fiscal agent fees, derived from the overall state-wide performance of school construction bonds and aggregation of the market. The centralization also provides the opportunity for the Division to promote energy efficiency in school construction and rehabilitation projects. As a result, schools have been able to install GSHP for their heating and air conditioning needs for about $4 per square foot of interior space to be served. At 2006 power costs, these systems can pay for themselves in seven to ten years, with average annual savings of about 20 per cent. Kentucky boasts that almost a quarter of its school buildings use GSHP and cooling systems. It has also enhanced the energy efficiency of 17 Energy Star-certified schools, showing an average saving of 45 per cent. These schools include retrofitting of existing buildings, some over 50 years old.

As energy consumption drops, the prospect for cost-effective solar photovoltaic installations in the Kentucky climate that could cover the remaining electricity needs pushes the plans towards net zero energy consumption buildings. As the Warren County school experience shows, however, getting to this standard is more than a technical matter. It involves attitude and behaviour change, community co-operation and staff training. While 40,000 square feet of solar collectors on the roof of the new school has provided the zero net energy result, that accomplishment would not have been possible without the prior investment in the structural envelope and the modification of management behaviours of the building users.

How Can We Make it Work Here?

At the start of this chapter we talked about the show *Dragons' Den*. A better analogy than dragons might be elephants or ghosts. Infrastructure has a long memory. Decisions made by our forefathers 100 years ago still haunt us now. Long ago decisions about which direction to orientate our houses now impact on the feasibility of installing a solar hot water heating system, while the lack of easy access to a back garden by heavy drilling equipment is a show stopper to installing ground source heat pumps at many properties. The dense cobweb of pipes and sewers and lines beneath our roads affect the viability of laying new DH networks, yet this might be the cheapest way of substantially reducing the energy use by Prashant's household. This is the view of staff in the local council and also of some of the councillors. Infrastructure has a long memory and it doesn't forgive bad decisions.

In this book we reject the idea that distant corporations should decide how much and where to invest in our energy infrastructure. Our system of financing energy infrastructure ought to equip local communities to take over its ownership, not just in a legal sense, but in the societal sense that reflects their preferences and trade-offs and provides local employment and uses local energy resources where possible. The Tea Party movement in the US and the idea of the Big Society in the UK are an articulation of this desire to recapture local decision-making. The funding and ownership of infrastructure is a large part of this re-conquest.

This chapter has used examples from Europe and North America to show how communities have established localized ownership of energy. They have raised finance, assigned responsibilities for risks and arranged for investors to be repaid. Where this has been made to work, it has brought profound, indeed transformational, change to the energy use of the community. The Danish deployment of DH and wind and Upper Austria's successful use of solar thermal, hydro and biomass-fired DH have changed patterns of employment and made their communities more energy secure, as well as reduced the emissions of greenhouse gases. By staging and planning their investments intelligently, energy prices have remained affordable too. These have been great success stories; many of the other examples have been qualified successes. The chapter excludes the many failures that we know have taken place too. We know of examples where money was allocated to projects and never spent, ESCOs that were established with huge political fanfare and massive hopes but which never completed a deal, and investors who speculatively sank millions into low-carbon infrastructure projects and lost their shirts. These set back community energy investment in the sponsor companies' minds for many years. We could be specific and name examples, but we wish to spare blushes.

The truth is that energy policy in most of Europe and North America is getting more things wrong than right. It is supporting high-cost micro-renewables ahead of lower-cost community renewables. Most of the licensing for new power plant that the market is seeking continues to be gas and to a lesser extent coal, despite their security of supply (in the case of gas) and carbon impacts. There is a desire for instant political gratification in energy policy that needs to demonstrate results within the two to three-year political and spending cycle, when the appropriate gestation period is much longer. Again the analogy of the mammalian elephant that spends 22 months in its mother's womb is better than the reptilian dragon, where eggs are spat out indiscriminately in the hope that one might flourish. There is also the excessive faith that competition between companies in liberalized markets, tweaked with a few incentives to encourage security of supply and renewables, sends the appropriate signal to investors.

Our examples show the precise opposite. Local communities need to have a plan, they need to articulate a vision of the building, street or neighbourhood they are trying to create and sell this vision to their people. Once that process is complete, they seek investment, Our stories confound any hope that there is a single solution; no two human settlements – or governments – are the same. Thus there will always be a replication problem. It may not be enough to know that 'they' did it. The question will remain, 'can WE?'. Kentucky schools have a unique financing system. Rotterdam represents a scale of operations – and some geological sequestration opportunities – that are not common to all local communities. Copenhagen's DH successes are the result of decades of national effort to promote the technologies. Vermont faces exceptionally high power costs relative to much of the rest of US. In Toronto David Miller diagnosed the problem as being disfiguring and inhospitable tower blocks that need renewing.

This chapter shows us there are a number of things we need to get right in order to make investment in sustainable community energy systems viable. Government must ensure low-cost capital is available to community energy developments:

- It can help reduce the cost of borrowing by lending its reputation through guaranteeing loans, or it can pay in capital.
- Local governments need to ensure that loans in discrete schemes are appropriately aggregated before going to the money markets to avoid excessive transaction costs.
- Projects should be actively refinanced to ensure that working capital is not tied up.

We believe that the beneficiary should ultimately pay for investment in community energy through higher energy or property charges. Costs should not be borne by the general taxpayers or the general energy consumer. Government capital grants should be paid for novel technologies or assist poor and vulnerable communities to access the community investment. Governments should commit to a level of grant many years ahead.

Investment plans need to be agreed locally – this allows wider benefits (employment, sector support, fuel security, affordability) to be given due weight and broadens the constituency of support. Local government has an important role to play in acting as anchor customer, and regulating and enforcing to ensure community decisions are respected.

It's important that the beneficiary, rather than the general energy customer, pays for improvements to their home or their local energy infrastructure. But people hate paying up front and some can't afford it; so we need to present the investment with minimum upfront costs, or show it increases the value of the house. Ideally we can use government's ability to borrow cheaply to access capital and then pay the costs of discharging the loans through savings on the energy costs. The US and the UK are developing ways of attaching the costs of improving homes to property taxes and energy bills respectively. This could be extended to recovering the cost of investment in community assets such as new networks.

Notes

1   Theo Paphitis is an entrepreneur and investor on the UK TV show, *Dragons' Den*. This is a regular refrain of his on the weekly show.
2   Creyts, J., Granade, H. and Ostrowski, K. (2010) 'US energy savings: Opportunities and challenges', *McKinsey Quarterly*, January, www.mckinseyquarterly.com/ US_energy_savings_Opportunities_and_challenges_2511.
3   European Commission (2010) *Energy 2020 A Strategy for Competitive, Sustainable and Secure Energy, SEC(2010) 1346*, www.ec.europa.eu/energy/strategies/2010/doc/ com(2010)0639.pdf.
4   Purple Market Research (2009) 'Solid wall insulation supply chain review – prepared from Energy Saving Trust and Energy Efficiency Partnerships for Home', www.eeph. org.uk/uploads/documents/partnership/SWI%20supply%20chain%20review%20 8%20May%2020091.pdf.
5   Holmes, I. (2010) 'Accelerating the transition to a low carbon economy', E3G, London, www.e3g.org/images/uploads/Accelerating_the_transition_to_a_low_carbon_ economy_The_case_for_a_Green_Infrastructure_Bank.pdf.
6   Department for Business, Enterprise & Regulatory Reform (2008) 'Heat: Call for evidence', http://webarchive.nationalarchives.gov.uk/+/http://www.berr.gov.uk/files/ file43609.pdf.
7   UNEP (2009) *Energy Efficiency And The Finance Sector*, Finance Initiative's Climate Change Working Group, http://ccsl.iccip.net/energy_efficiency.pdf, p4.
8   Peter Meyer, who helped with the primary research for this book, is currently under contract to the US Department of Energy to help those recipients use the funds more rapidly and effectively. These observations come from his field experience and interaction with other technical assistance providers to the EECBG grantees. The failure of the fund recipients to spend them rapidly has become a political problem, as the monies were intended to stimulate the economy in the face of the recent recession.
9   An English translation of the German feed-in-tariff regulation can be found at www. erneuerbare-energien.de/files/pdfs/allgemein/application/pdf/eeg_2009_en.pdf. Part 4 discusses the equalization payments to ensure that the incidence of tax is according to the amount of electricity sold by each electricity supplier.

10 Hinnells, P. and Rezessy, S. (2006) 'Liberating the power of Energy Services and ESCOs in a liberalised energy market', Environmental Change Unit, Oxford University, www. eci.ox.ac.uk/research/energy/downloads/bmt-report3.pdf.

11 Interviews with Siemens representative, 2008, and also information from the National Association of Energy Service Companies website, www.naesco.org.

12 Holmes, I. and Mabey, N. (2010) 'Accelerating the transition to a low carbon economy: The case for a Green Infrastructure Bank', E3G, London, www.e3g.org.

13 Transform UK & E3G (2010) 'Written evidence to the Environment Audit Committee on the Green Investment Bank', www.publications.parliament.uk/pa/cm201011/cmselect/cmenvaud/memo/greeninvest/wrev35.htm.

14 Personal communication with staff from B&Q Pay As You Save pilot. Workshop on Pay As You Save, 10 July 2010. Prashant is on the Advisory Panel for the Pay As You Save pilots and was involved in the original development of the Distribution Network Operator model – the not-very-sexily-titled original name of the Green Deal scheme.

15 Vaze, P. and Mayo, E. (2009) 'A new energy infrastructure', Consumer Focus, London, www.consumerfocus.org.uk/assets/1/files/2009/06/A-New-Energy-Infrastructure2.pdf

16 See www.youtube.com/watch?v=XCSsH9Jl_Mo.

17 Morrison Park Advisors (2010) 'Tower renewal financing options', www.toronto.ca/city_manager/pdf/tr_financing_options_report.pdf.

18 Morrison Park Advisors (2010) 'Tower renewal financing options', www.toronto.ca/city_manager/pdf/tr_financing_options_report.pdf.

19 Egger, C., Ohlinger, C., Auinger, B., Brandstatter, B. and Dell, G. (2010) 'How Upper Austria became the world's leading solar thermal market', www.esv.or.at/fileadmin/redakteure/ESV/Info_und_Service/Publikationen/Solar-publ-eu.pdf.

20 Kentucky Department of Education (2009) 'KABC Green Site', www.energyeducation.com/Portals/0/Warren_Co_KY_ASE_Andromeda_Awards_062209.pdf.

Chapter 7

Increasing the Price of Energy and Helping Renewables Without Punishing the Poor

You can have clean energy or you can have cheap energy. You can't yet have cheap clean energy.
Ed Miliband (2010), UK's former Energy and Climate Change Secretary[1]

To make the transformation needed to control climate change and increase energy security, the price of fossil fuels needs to increase substantially. As Ed Miliband pointed out, developing clean energy is more expensive than dirty energy. Phasing out unsustainable energy sources will ultimately mean cleaner air, improved health and reduced medical bills. Relying on the wind, the sun, the waves and the tides will mean that most countries can get much more of their energy from within their own borders or from friendly countries. Oil wars such as Iraq can be consigned to history.

The leading and most authoritative voice on the economics of climate change is Nicholas Stern. Stern has worked at World Bank and in the UK Treasury. While at the Treasury, he led a team compiling *The Economics of Climate Change*, published in 2007.[2] The review is exhaustive (700 pages long) and concludes that the cost of controlling climate change is less than the cost of not controlling it. Despite being a civil servant, Stern subsequently showed a politician's ability to coin a soundbite by describing climate change as:

the greatest market failure the world has seen. The evidence on the seriousness of the risks from inaction or delayed action is now overwhelming ... The problem of climate change involves a fundamental failure of markets: those who damage others by emitting greenhouse gases generally do not pay.[3]

Stern is right that acting will be cheaper than not acting. But this does not mean that it will be cheap. This chapter is about how taxes and emission trading schemes (ETSs) can be used to find and direct resources to stimulate sustainable energy use. Our ambition is merely to ensure that the price point is in the right direction and is stable. We use examples drawn from European nations and US states instead of community-level initiatives. Policy to tax energy and create price subsidies is usually the preserve of these tiers of government. It makes no sense for counties and cities to set their own energy tax rates as businesses would relocate to avoid high cost areas.

This chapter first looks at why we *need* taxes and reminds us that raising taxes is not without political difficulties. To be fair and seen as acceptable, the taxes need to be smart. Second, it reviews the role *energy taxes* and cap-and-trade schemes have played in Europe and US. Third, the chapter looks at how *earmarked revenues* raised from these 'taxes' are used to incentivize renewables and energy efficiency to help us grope our way to stimulating and financing community energy resources. The chapter finishes by looking at the *structure of energy tariffs*.

Why Do We Need Energy Taxes? And What Do We Mean by Smart Taxation Policies?

Consumer advocates want rate reductions, not rate hikes ... TURN is fighting at the CPUC, [utilities] asks for a historically high $4 billion increase, which would surely hit consumers hard.

The Utilities Reform Network, California[4]

I love paying tax ... it's what separates us from tortoises and grapefruits. We pay tax to help the ones that aren't as fortunate as us.

Ian McMillan (2010)[5]

Consuming electricity is like taking different things off supermarket shelves, without knowing their individual price, and then pushing the cart up to the cashier and waiting for the bill to come out.

Fabrice Haïat, Vizelia[6]

The above quotes give some indication of the polarization of views about energy prices. Consumer groups such as TURN argue that price rises sanctioned by the energy regulators are soft on the utilities and unfair to consumers, vulnerable consumers in particular. But as Ian McMillan reminds us, seen another way, a fair tax is also a manifestation of our humanity – a statement that as a society we are in it together – resources need to be taken from one person and given to another. And the absence of timely and comprehensible information on gas and electricity bills means that apart from the small proportion of the population that are energy anoraks or budgeting obsessives, hardly anyone understands how much they pay for their energy.

So which is it? Do we pay too much or too little for our energy, or are we totally clueless and therefore indifferent to price signals? In reality it probably depends on the customer. People hate new taxes. The event that triggered the establishment of the US was a tax on tea. The Tea Act allowed the monopoly East India Company to impose what was perceived to be an unfair tax levied on tea being imported into the US straight from India, for the benefit of its British share-holders and the British government. American citizens were being taxed but had no representation in the British Parliament. That event is still an effective rallying cry in contemporary US politics two centuries later. India's freedom movement used the salt tax (yes, you've guessed it, imposed originally by the East India company in 1835![7]) to inspire civil disaffection with British rule through the late 19th and early 20th centuries. Mahatma Gandhi's Satyagraha, or salt march, in 1930 marked the start of his civil disobedience movement against the Raj. Gandhi was a shrewd politician – he explained why he selected salt saying, 'Next to air and water, salt is perhaps the greatest necessity of life'.

In the previous chapter we talked about the investment needed to make our energy systems more sustainable. We argued that it should be paid largely from energy customers benefiting from the investment. But some of it must also come from general government revenue. We believe that carbon and energy taxes should be part of the general government revenue; the same as income taxes and consumption taxes. Energy and carbon taxes are capable of raising substantial sums of money (and already do in Denmark, Germany and Sweden). We recognize the politics of increasing energy costs through taxation are fraught, particularly in Anglo-Saxon countries, but we think this is a political battle that needs to happen. We need to set high energy and carbon taxes to encourage individuals and organizations to use energy efficiently, and to raise revenue.

Large public deficits make it unlikely that most of the energy transition will be paid for out of public funds. Table 7.1 shows how much revenue could be raised if taxes on gas and electricity were raised to the same level as those presently paid by Danish households (Danish industry pays lower energy taxes to prevent hurting their competitiveness with businesses in other countries). Energy taxes can raise big money. The €58 billion from a UK energy tax would be enough to pay for around a third of the structural budget deficit that the economy faces and which is necessitating the dramatic changes in taxation and public spending the country is going through. The €540 billion per year from the US energy tax is about the same as the deficit in the 2008 budget (but nowhere near enough to pay for the $1.4 trillion deficit in 2009).

A small share of the proceeds from energy taxes would be sufficient for government to make its contribution to the investment in the sustainable community energy systems. Intelligent environmental taxes can do good. The idea behind green taxes is that they will discourage wasteful or excessive use of resources. They have been used quite successfully in other environmental issues including

Table 7.1 *Potential revenues from energy taxes levied at the same level as household energy taxes in Denmark and current levels of household energy tax across selected countries*

	Potential revenue from Danish-style energy tax (€ billion)		Current Level of household energy tax in 2009 €/kWh	
	Electricity	Gas	Electricity	Gas
US	337	201	0	0
Germany	16	23	0.06	0.006
Netherlands	8	6	0.02	0.018
UK	31	27	0	0
Denmark			0.09	0.029

Source: author's calculations based on Eurostat and US Energy Information Administration data

transport, waste disposal and the quarrying of aggregates. The use of taxes to reduce greenhouse gas emissions has been less successful in the past, partly because they have been the cause of much more extensive lobbying by powerful and rich sectors that oppose any reduction in demand for their products, and partly because access to energy for warmth is essential to human survival, unlike, to take an example where taxes have been used successfully, plastic carrier bags in Ireland.

The idea that energy taxes can do good might seem a bizarre statement for Prashant to put his name to. He works for a consumer organization that chides energy companies when they increase energy prices. But consider this: in the UK, the richer half of the population spends just 2 per cent of income on domestic energy (the same share of income they spend on package holidays). This is even though they use around one and a half times the energy as the poorer half. Energy taxes – including taxes on domestic energy – are needed to moderate demand.

But this must be done as part of a package of policies, which ensures that poor people are not made to suffer even more than they already do, and that businesses that are exposed to international competition do not simply relocate to other countries. In the UK, 17 per cent of the population (4.5 million households) spend more than a tenth of their income on warmth and electricity.[8] In Scotland the figure is 27 per cent.[9] The winter death rate is 19 per cent higher than the summer death rate, in part due to ill health caused by colder damper conditions in people's homes. Across Europe there are several other countries with similar sorry statistics on high levels of fuel poverty, often countries with moderate climates, like Ireland and Spain.[10]

Figure 7.1 shows the strong relationship between spending on energy (broken into ten equal groups according to how much they spend) and income (broken into ten equal groups according to how much they earn) across many thousands of households that were quizzed between 2004 and 2007. The width of each bubble shows the number of households that lie in one of the hundred combinations

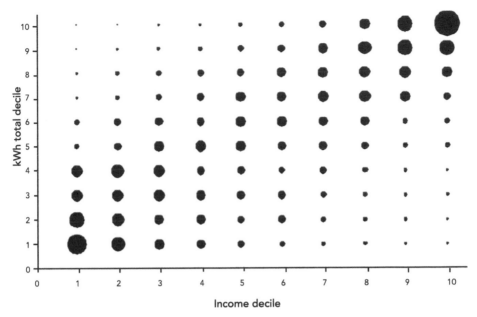

Figure 7.1 *Distribution of spending on energy across UK households for different incomes*

Source: Centre for Sustainable Energy and Association for the Conservation of Energy[11]

of energy spend and income. The distribution shows there is a linear relationship between income and spending on energy – the diagonal line of big bubbles. But there are a small number of households in the top left of the graph. These are the poor households who consume a lot of energy. High energy prices will hurt these people and we need to consider how to protect them.

Fuel poverty is, of course, just one aspect of poverty. It is much less widespread in Scandinavian countries (which have much colder winters),[12] The Netherlands and Germany than in the UK or US, partly because these countries have much less unequal distributions of income and wealth. This is beyond the scope of this book, but the relative lack of fuel poverty in these countries is also because they have sensible energy policies and decent buildings. We briefly examine the policies and programmes they have used to improve energy efficiency and avoid fuel poverty.

Energy and Carbon Taxes in Europe and the US

Jacques Delors spent much of his period as president of the European Commission arguing for an EU carbon/energy tax. But he failed to get the member states to agree to introduce one. Taxation measures require unanimity in the Council of

Ministers, so any country, even tiny Luxembourg, can theoretically prevent their adoption. In fact many member states, including the UK, opposed the Delors proposal on subsidiarity grounds – the argument that taxes are for national governments to decide, not the EU. Other countries less opposed to EU integration, such as Germany, opposed the Delors proposals because it would have damaged their coal industries. (The two largest German energy utilities, E.ON and RWE, both have very large coal operations and are well connected with German politicians.) Instead of energy taxes, the EU Commission has gone down the ETS route, which restricts the amount of CO_2 that can be emitted by large industrial and energy-generating sources, thereby causing the market to generate a price for the right to emit this CO_2. At a theoretical level a carbon tax and the EU-ETS can be equivalent.

The only EU measure (Council Directive 2003/96/EC) on energy taxation is a required minimum level of tax on petrol and diesel, but this is too low to promote energy efficiency significantly. It is currently set at a rate much lower than actually levied by countries in northwest Europe. There has been some progress on carbon and energy taxation in European countries, notably Scandinavia and The Netherlands. Due to taxation, there are considerable differences in the energy prices paid by consumers in different European countries. Households in Denmark and Sweden pay by far the highest gas prices, about 50 per cent more than the next highest in The Netherlands. In 2009 British households paid the lowest gas price of any EU member state. Denmark has the most expensive domestic

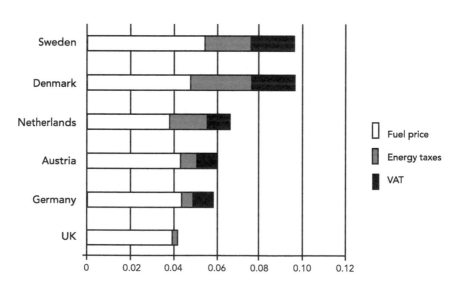

Figure 7.2 *Natural gas prices for domestic consumers in 2009 (€/kWh)*

Source: Eurostat[13]

electricity tariff in the world. German domestic electricity tariffs are around two-thirds of those in Denmark. The UK does not raise revenue from energy taxes on households and charges an abnormally low rate of value added tax (VAT) (see Figures 7.2 and 7.3).

The tax on domestic electricity in Denmark of 9c/kWh is equivalent to €150/tCO$_2$, around ten times the current price of carbon in the European emissions market. In Germany and Scandinavia, revenue from carbon and energy taxes was used to reduce taxes on income or employment. This was often referred to as an environmental tax reform (ETR) and was done to deliver a 'double dividend' – lower levels of pollution and higher employment. Some revenue in Sweden has been used for energy efficiency work. There is an administrative/bureaucratic objection to a green tax shift in the UK: the Treasury objects to allocating (which it calls 'hypothecating') particular revenue streams to particular spending streams or reductions of other taxes.

Figures 7.2 and 7.3 set out the overall levels of energy taxes in 2009. Interest in taxes has waxed and waned depending on political ideology, pressures from industry and consumer lobbies. The manner in which taxes have been applied varies between country and over time: sometimes exempting particular sectors or according to the carbon, sulphur or nitrogen content of the energy source to address climate change or acid rain. Box 7.1 gives a potted history of environmental taxes in a number of European countries.

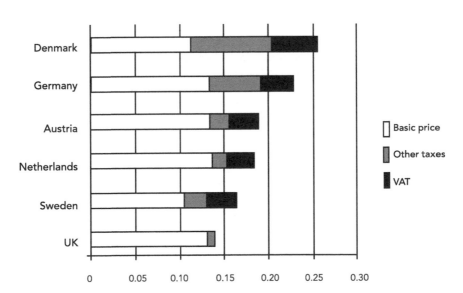

Figure 7.3 *Average electricity prices for domestic consumers in 2009 (€/kWh)*
Source: Eurostat[14]

Box 7.1 Green taxation in selected countries in northwest Europe

Norway

- 1970: tax on sulphur in mineral oil. First environmental tax.
- 1991: CO_2 tax on mineral oil products, later extended to all fossil fuels. Different fuels pay different rates: in 2009, the highest rate – mineral oil – was the equivalent of €27/tonne.
- 2007: tax on nitrogen emissions.

Denmark

- 1977: electricity tax. The aim was not environmental, but to reduce the balance of payments deficit resulting from the import of oil products and to increase the use of natural gas.
- 1992: carbon tax. This was dependent on the content of CO_2 in the fuel, and the highest rate was the equivalent of €13.5/tonne. However, the aim was not to increase the price of fossil fuels, so the energy tax was lowered.
- 1996: sulphur tax.
- 2010: nitrogen tax.

Overall, Denmark has the highest rate of energy taxation of any EU country. However, the Danish approach to heavy industry has been rather weak: firms using more than 15 million kWh a year of electricity do not pay energy tax on the amount above that level, and energy-intensive industries are fully exempt from paying the CO_2 tax. The Danish tax shift has involved 2.3 per cent of GDP or about 5 per cent of the tax raised by the government.

The Netherlands

- 1990: carbon tax, changed into a carbon/energy tax in 1992. This mainly affected large energy users.
- 1996: carbon/energy tax to target small-scale energy consumers. All households and about 95 per cent of Dutch companies are covered by this tax.
- 2008: carbon tax on coal (which provides around a tenth of Dutch energy) at €26 per tonne.

There are few exemptions to Dutch energy taxes, although ironically for a policy to tackle the greenhouse effect there was at first no tax on natural gas used to heat actual greenhouses, notoriously high energy users. Revenue from Dutch energy taxes was used to reduce personal and corporate income tax and also to offer accelerated depreciation for energy investments. Some is distributed to households as an Energy Premium (see section on fuel poverty below). The Dutch ETR involved around 0.7 per cent of GDP.

Sweden

- 1991: sulphur dioxide tax.
- 1991: carbon tax. This is the major form of energy taxation in Sweden and has been successively increased since 1991. In 2009, it was over €40/

tonne. However, industries are required to pay only 50 per cent of the tax, so in practice the Swedish rate is similar to the Finnish one. Also, no tax is applied to fuels used for electricity generation, so the main effect has been on promoting renewable heat.

• 1992: nitrogen dioxide tax.

Some of the revenue from the various energy taxes was used to reduce income tax. Overall, 4.6 per cent of Swedish GDP was involved in this tax shift.

Germany

• 1999: existing energy taxes, particularly on transport fuels, were increased and a new electricity tax introduced. The government pre-announced that the electricity tax (and petrol and diesel taxes) would increase every year: this was done annually from 1999 to 2003.

Most of the revenue from the green taxes has been used to reduce employer and employee social security contributions, though a small proportion has been used to support renewables and renovate existing buildings to make them more energy efficient. This tax shift amounted to only about 0.6 per cent of GDP. The tax rates are low and there are numerous exemptions and reduced rates. One example: businesses that pay 20 per cent more in the new green tax than they save in lower social security contributions are partially reimbursed by the federal government. Germany has an energy tax, not a carbon tax. Renewables are exempt and good quality combined heat and power (CHP) pays a lower rate. However, there is no formal link between tax rate and carbon content. Most strikingly, brown coal and hard coal, which are very high in carbon content, are exempt from the energy tax.[15]

UK

• 1993: VAT on domestic energy. The then Conservative government introduced this at 8 per cent. It then tried to increase it to 15 per cent, but failed to get this through Parliament. After Labour came to power in 1997 it reduced the rate to 5 per cent (the lowest rate permitted under EU rules).
• 2001: Climate Change Levy. This is a tax on energy use in industry, commerce and the public sectors, not levied on households. Despite its name, the Climate Change Levy is not a carbon tax – the same rates are payable on coal, gas and nuclear electricity, though CHP and renewables pay lower rates. As explained in Chapter 3, this was because the Labour Party was not prepared to alienate its supporters in the coal industry. The Conservative/Liberal Democrat coalition government, which is not close to the coal industry, says that it will turn the Climate Change Levy into a carbon tax.

Rates are low: 0.15 pence/kWh for gas, 0.44 pence/kWh for electricity and 0.12 pence/kWh for any heating fuels. The main impact of the Climate Change Levy has been to focus companies' attention on energy efficiency. Many sectors entered into agreements to improve energy efficiency in return for a reduction in the rate of tax. The revenue from the Climate Change Levy was used to reduce employers' national insurance contributions (though they subsequently went back up), with a small proportion used to provide support for energy efficiency and renewable energy. This tax shift represented only 0.06 per cent of UK GDP.

So how effective have these taxes been at reducing emissions? It is of course difficult to be certain about the impact of an energy tax, as there are so many other things changing at the same time including underlying fuel prices, economic growth and change in industrial structure and weather. But assessments show that green taxes can be effective, Speck (2009)[16] concludes:

- Norway's carbon tax has led to a reduction of 21 per cent in CO_2 emissions from power plants. The tax is said to have reduced total Norwegian carbon emissions by 2 per cent. Carbon emissions per unit of GDP have reduced by 12 per cent.
- Denmark's carbon and energy taxes have reduced emissions from affected sectors by 6 per cent.
- In Sweden, emissions would have been 20 per cent higher than 1990 levels without the carbon and energy tax.
- The Netherlands' emissions are 3.5 per cent lower than they would have been without the carbon and energy tax.
- Germany's CO_2 emissions were 2–3 per cent lower by 2005 than they would have been without the carbon/energy tax.

So taxes are an important part of the policy mix. But none of the emissions reductions is nearly large enough to control climate change. And none of these governments thinks that once a tax is implemented the issue can be 'left to the market'. In addition to market mechanisms, strong regulations are essential. The Danish experience shows better than anywhere else what can be achieved through a combination of taxes and regulation.

A carbon tax is one way to try to reduce CO_2 emissions. An alternative is a cap-and-trade system. Under this approach, a public authority decides on an overall level of acceptable pollution or 'cap' and allocates or sells off permits up to the cap. Firms can buy and sell their permits if they have surpluses or deficits. Cap-and-trade was used by the New England states in the US to control acid rain pollution. This was a successful policy, even though acid rain gas emission permits would not be the first choice as exemplar for an ETS since the permits are not really equivalent. Location as well as quantity matter – a fact the trading system neglects. Greenhouse gas emissions are better for trading because it makes no difference where they are emitted. Cap-and-trade is theoretically better than taxes if the objective is to attain certainty about the level of greenhouse gas emissions; taxes are better if industries need the price of the permits to be certain, to remain internationally competitive. This is a theoretical benefit rather than an actual benefit; much of this certainty can be eroded in the precise design of any scheme. What if political lobbying sets the cap too high? Should companies be allowed to buy CO_2 savings from projects outside the scheme? What happens if firms start replacing energy sources in the scheme, for example using power stations with small energy sources excluded from the scheme, such as domestic diesel generators?

Having failed with Delors's carbon tax, the European Commission pushed for the ETS. This was introduced in 2005. It was a start, but has so far delivered few results. National governments insisted that they, not the Commission, were the public authorities to decide the maximum permitted level of pollution, and allocated far too many permits, largely concentrated in the hands of the large power companies. The price per tonne of carbon has been far too low and too volatile to incentivize investment in sustainable energy. From 2012, the Commission will control the total number of permits EU-wide, which should improve its ability to create genuine scarcity in the market. But it will do so on the basis of a calculated 'baseline' – the amount of pollution predicted under current policies. This baseline needs to fully take account of the economic recession, which has led to a significant fall in carbon emissions. The recession is good for the climate, but not for the move to a low-carbon economy, and not for the millions who have lost jobs and income.

As well as the Commission deciding the total number of permits, EU governments have agreed that permits will be auctioned, rather than handed out without payment as they have been in the past. This was strongly opposed by the energy industry, including the German utilities, which condemned it as anti-coal (which is the entire point of the ETS). But Angela Merkel was sufficiently determined to overcome the opposition. Auctioning of permits will produce substantial revenue, some of which has already been allocated to low-carbon energy sources (CCS and new renewables). If carbon is more scarce after 2012, it will make renewable energy more cost-competitive than fossil fuel energy.

One option to improve the effectiveness of the ETS would be to introduce a floor price. The UK Conservative/Liberal Democrat government is committed to this (and say that making the Climate Change Levy an upstream carbon tax will in practice do this). However, at the time of writing, the government has yet to say what the tax rate/floor price will be. The higher the rate, the more effective it will be at stimulating low-carbon options but the greater the increase in energy tariffs. The government says that rebates will be offered to reduce the impact on some sectors, particularly the domestic sector.

The US does not have a national cap-and-trade system. The Obama Administration tried to introduce one, but failed to get it through Congress. There is, however, a regional scheme, the Northeast Regional Greenhouse Gas Initiative (RGGI), which commenced in 2009 and another nascent scheme, the Western Climate Initiative (WCI), which is scheduled to start in 2012. Ten states including New England and New York are participating in the RGGI. Carbon emissions from power stations larger than 25MW are now capped. (The threshold in the EU scheme is 20MW.) Carbon is auctioned every three months. In the first year around 85 million tonnes of CO_2 were sold at around $3/t$CO_2$. In comparison, the EU scheme covers around 2000 million tonnes of CO_2 and prices are around €15/tCO_2.

Earmarking Revenues for Energy Efficiency and Renewables

Even supporters of carbon taxes and ETSs will accept they haven't brought about much investment in alternatives. The tax has been too low and the price of carbon too volatile to bring on renewable energy or real investment in energy efficiency. David Farnsworth of the think-tank RAPonline told us that their modelling of the RGGI and WCI suggested the higher price of energy only accounted for a fifth of the reduction in energy use. The rest would be down to what RAPonline call complementary measures.

The energy market is not great for renewable energy for three reasons. None of these are inherent problems with renewables; the problems are with the market. First, the constantly changing price of electricity and heating fuels is linked to the global price of gas and other fossil fuels. This bears no relationship to the changing costs faced by renewables, which arise from servicing their debts (interest risk) and the costs of feedstocks such as biomass (agricultural price risks). This means their costs and their revenues could be moving in totally the wrong directions. Second, the renewable developer needs to take on long-term debt and so needs a long-term certainty of getting a decent price for the power and heat. But vertically integrated suppliers don't want to enter such contracts with independent generators. They would rather contract with their own subsidiaries. Third, the energy market has developed around the needs of large generators so electricity and gas is traded in large packets, with long-term contracts with financial intermediaries to strip out risks.

So despite the taxes and cap-and-trade systems, clean energy remains more expensive than fossil fuels and disadvantaged in the marketplace, so complementary measures are necessary. As well as providing direct resources for energy companies or customers to install measures, they have also stimulated innovation and the development of skills among installers.

Many countries use a FiT to support renewables. Renewables are paid a fixed price for supplying electricity to the grid, decided by a public authority. The contract lasts 20 years and is regarded as reliable enough to go to a bank to raise finance with. A FiT can also be used to support CHP using fossil fuels like gas, as it is in Denmark, or for renewable gas.

Energy efficiency installation targets are also financed by levies on energy bills. This is explicit in many US states; the utility regulator monitors how much energy companies spent on energy efficiency programmes, and how much they achieve in terms of installing low-energy lighting. In countries such as the UK the levy is implicit; the requirement to install energy efficiency is imposed on electricity and gas suppliers but they are free to do this as cheaply as they can. Levies to support renewables and energy efficiency have increased UK domestic gas prices by around £80 per household – around 5 per cent of the bill.[17]

Box 7.2 FiTs in Germany – waste of money or renaissance for renewables?

The Feed-in Law was introduced in Germany in 1990.[18] This beautifully short piece of legislation (it is just one page long) mandated that energy companies had to purchase electricity produced from small hydropower, wind energy, solar energy, landfill gas, sewage gas or biomass. The price paid was set in the legislation at between 60 and 90 per cent of the historic retail price of electricity, depending on the technology. This legislation was hugely successful in stimulating wind, since wind generators were guaranteed access to the electricity market, the price set at 90 per cent of the average retail price high was high enough for the technology to flourish, and the price paid was stable, unlike wholesale prices, allowing the developer to raise finance from banks. As a further twist, energy companies were not allowed to use the FiT (though this restriction was relaxed over time), stimulating new entrants into the power market, especially bottom-up community developments. The cost incurred in buying this electricity was to be shared between energy companies.

The legislation underwent a major restructure in 2000 when the Renewable Energy Act was passed. It addresses all the difficulties faced by renewables in the energy market. It detached the sale price of renewable electricity from the market price (which was really a reflection of fossil fuel prices) and linked each renewable technology to the cost of generation, *but in return for this certainty of making a reasonable profit* electricity prices declined annually to force technology to become more cost-effective, or pack up shop. These changes really got the industry going.

In 2009 the tariffs for different technologies were: large wind €0.09/kWh, offshore wind €0.13/kWh, small hydro €0.07/kWh to €0.13/kWh and biomass €0.08/kWh to €0.11/kWh. All of these charges are very similar (often lower) than the price retail customers pay for electricity, even before taxes and VAT (see Figure 7.3) and are only a little higher than the price received for large fossil generators (so-called 'grid parity'). The fossil fuel generators would argue (with some justification) that this is too flattering a comparison for wind since it is not being punished for its unreliability; fossil fuel has to pick up the pieces when the wind stops blowing.

The main exception is the tariff for solar photovoltaic: free-standing plant is paid €0.29/kWh. This is still more than the retail price of electricity in Germany. The tariff for small rooftop solar photovoltaic is more generous still. The policy has been criticized as the high rates paid for solar have been exploited to the detriment of customers that pay the generous subsidies. But the subsidies have greatly brought down the cost of the technology. The solar photovoltaic rates are now so attractive that farmers are building large solar farms on their land – greatly increasing the burden on electricity customers. The price paid for free-standing solar photovoltaic (often on arable land) was cut by 25 per cent in 2010. This sharp rate of decline worries the solar photovoltaic manufacturers and installers; innovations are still bringing the price of modules down, but costs are not coming down as fast as necessary.

Has the German FiT been a success? We think so. The real story has been the take-off of the renewables industry and wind in particular. Renewable electricity now accounts for 8.7 per cent of electricity used in Germany. Wind and biomass account for 74 per cent of this, with photovoltaic just 7 per cent.[19] In 2008 the FiT cost customers an extra €4.5 billion, but avoided the use of €2.7 billion in imported fuel and electricity.[20] The federal government also argues there are substantial further savings from the energy system as a whole.

The FiT has been the main renewables policy tool in Germany (see Box 7.2). The level at which a FiT is paid normally varies according to technology and size of plant. Technologies that are near-to-market, such as wind, are paid much less per unit of electricity than are those further from market like solar photovoltaic. (However, wind is still paid more than is electricity from fossil fuel or nuclear power stations.) Also smaller systems, such as those mounted on home rooftops, are paid more than large commercial installations. The tariffs also decrease every year to encourage innovation. It's a clever design and the policy has been very successful at hugely increasing the amount of photovoltaics installed in Germany.

For comparison, in Spain tariffs were set at a maximum of €0.44/kWh for solar. In California, the rate started at $0.39/kWh (€0.27) but is being reduced quickly. The tariff for wind in Germany is much less generous – in 2009, it was €0.092/kWh. This is broadly the same as in Spain. Critics of FiT argue the policy is perverse because it *rewards failure* – wind is being paid less because it is more efficient and does not need so much support.

In Germany, the costs of paying for the FiT are met through a cost-sharing mechanism for all electricity end-users, which means that all users' bills increase. The impact of this has been approximately €16 per year for every household.

The alternative policy approach to the FiT is the Renewables Obligation (UK) or Renewables Portfolio Standard (RPS) (US). This imposes on energy companies an obligation to sell a specified percentage of the electricity from renewables. In the UK, this has risen from 5.4 per cent in 2001 (when the Renewables Obligation was introduced) to 11 per cent in 2010/2011. In the US, the percentage target for the RPS is set at the state level. In California, energy companies have to increase their share by 1 per cent a year and achieve 20 per cent by 2010 and 33 per cent by 2020.[21]

In the UK, there was a fierce debate from 2001 to 2008 about the merits of the Renewables Obligation versus the FiT. This led to the worst of both worlds for renewables developers. The FiT delivers guaranteed income and so reduces the cost of capital for developers needing to borrow to invest and construct, compared to the obligation approach where income is uncertain. The Renewables Obligation was meant to deliver guaranteed output. But the target has never been met – putting the lie to this idea. So if starting from scratch, a FiT approach is economically better than the obligation approach. (The Renewables Obligation has itself morphed many times since it was first introduced – introducing different payments according to technology, minimum guaranteed prices for the certificates, and disentangling the revenue the developer receives from the market price of electricity. The RO has ended up being a FiT in all but name ... but with more paperwork.)

Whatever the policy, it has to be stable so developers can predict how much the support will be worth when the facility is complete and their subsidy is fixed. There might be an interval of several years between a project being mooted and it being completed. The Renewables Obligation has been amended seven times since 2002, giving rise to huge regulatory uncertainty. Arguably this matters more than the structural differences between an obligation and FiT.

So those wishing to develop renewables in the UK had to contend with a higher cost of capital due first to the obligation approach rather than a FiT, and second to regulatory uncertainty. This was one reason – though far from the only one – why the UK did so badly on renewables for so long (though a greater barrier to renewable development in the UK has been the land-use planning system).

In fact, there was never any need to decide between the Renewables Obligation and the FiT. It is possible and sensible to have both. Several US states have both the RPS and net metering, which requires the electricity network operator to pay for renewable electricity exported to the grid and so is similar to a FiT (though the rate is the same as for 'brown' electricity and so is not nearly as remunerative in promoting renewables).

In 2008 the UK government accepted that there was no need to abolish the Renewables Obligation. Instead, it introduced a FiT for anything up to 5MW. This was intended to increase the uptake of micro-renewables – photovoltaics or wind turbines on roofs. But 5MW is quite a high threshold, equivalent to two large wind turbines, and was chosen to encourage the expansion of community wind farms. Pro-FiT campaigners had argued that the Renewables Obligation is too complex a system for communities to master, though some communities have managed it. Community-owned wind farms are more widespread in Scotland than in England or Wales, partly because the Scottish government has sent officials out to communities expressing an interest, to help them fill in the forms.

The UK will be bringing out a new support mechanism, the Renewable Heat Incentive, for renewable heat technologies such as solar hot water, biomass and heat pumps. It will also support CHP that uses renewable heat sources. This will be a FiT-style mechanism for technologies where it is practical to measure the amount of heat produced.

As well as financial support for renewables via the RPS and FiT, renewables can be subsidized with grants. Fifteen US states and the District of Columbia have a public benefit fund that sets a charge for customers and spends the money on renewable energy and/or energy efficiency projects.

Levies are also used to support energy efficiency. After the 1970s oil shocks, many US public authorities introduced demand-side management (DSM) programmes, requiring utilities to spend money on energy efficiency programmes. In 1993, the year when DSM reached its peak, 447 US utilities were required to implement DSM programmes. They spent $3.2 billion that year. Seven European countries place energy efficiency obligations on energy companies: Belgium (Flanders region), France, Italy, the UK, Ireland, The Netherlands and Denmark. There are substantial penalties for companies that do not fulfil their obligations, but in practice penalties are rare as the targets are always met. In Flanders and the UK, the energy companies are required to ensure that there are also savings for low-income households.

The UK first imposed energy efficiency obligations, which require the saving of a certain amount of energy, on energy companies in 1994. These have become much more ambitious and have been responsible for a high proportion of all the UK's retrofit energy efficiency activity over the past 15 years. The current Carbon Emissions Reduction Target (CERT) requires large domestic energy suppliers to install around 100,000 substantial measures such as cavity wall insulation, loft insulation and more experimental measures such as solid wall insulation.

Big companies often 'met' these obligations in ineffective ways, such as by distributing millions of low-energy light bulbs, most of which ended up in cupboards rather than lamps (though the government eventually worked this out and changed the rules so this approach no longer counted as saving energy). Companies did implement some schemes trying to persuade people to save energy, but these suffered from quite understandable credibility problems, akin to getting an estate agent to discourage people from buying a home.

The best projects were those where companies worked with local councils. In 2004 British Gas began working with Braintree District Council, offering homeowners cavity wall and loft insulation. The best aspect of the scheme was that those who took up the insulation were given council tax rebates, thus combining a common-sense approach with a politically astute one. The scheme led to 1200 installations completed in the first five years. Over three-quarters of those who took up the insulation said that they would not have had the work done if the council tax rebate was not on offer. Braintree Council also used this opportunity to promote funding available under the Warm Front scheme. This model of local partnership has proved so successful it is now available in more than 60 local authorities.[22]

Kirklees Council and Scottish Power have had an even better approach: an area-by-area insulation project, which reduces costs and leads to 'neighbour pressure' to participate. A quarter of properties in the areas included have been insulated: 36,000 lofts insulated and 17,000 cavity walls filled. This project is reported to have created 129 local jobs and led to £7.8 million worth of fuel bill savings per year. Following this lead, the Welsh government announced an area-based whole-house programme for energy efficiency.

Energy Prices, Vulnerable Customers and Business Competitiveness

For most people, not being able to keep warm in winter is a matter of discomfort but for some it's a matter of life and death. There is no international definition of what constitutes fuel poverty. The UK definition is when more than 10 per cent of disposable income needs to be spent on fuel bills. Others measure it as the difference between summer and winter death rates or the number of consumers who are in arrears with fuel bill payments. The UK has more 'excess winter deaths' than other northern European countries with similar or colder climates, as the Table 7.2 shows.

Spain and Portugal have higher winter death rates than the UK despite being much warmer, which underlines the fact that fuel poverty is primarily a result of poverty and homes that are not insulated properly. Apart from ending poverty – highly desirable but not easy to achieve – we can pay vulnerable people cash to help them in winter.

The UK government gives out £200 a year to all those aged over 60 (£250 to the over 80s). This costs £2.7 billion a year. The payment is called the Winter Fuel Payment. It is supposedly to help them pay their winter fuel bills. But it is not means tested, and not all those over 60 are poor, and no one is required to spend it on energy (you can spend the winter months on a yacht in the Bahamas and still receive it). This is a waste of money. The UK government also pays a Cold Weather Payment to households with vulnerable people on means-tested

Table 7.2 Excess winter mortality rates in northern European countries, averages, 1988–1997

Country	Excess winter mortality as per cent increase over non-winter deaths
Austria	14
Belgium	13
Denmark	12
Finland	10
France	13
Germany	11
Republic of Ireland	21
Netherlands	11
England	19
Scotland	16
Wales	17
Mean	16

Source: Healy (2003)[23]

benefits when the temperature drops to below zero for a whole week. The benefit is worth £25 per week. This is a well-targeted benefit that is added automatically to qualifying people's pensions.

In most Scandinavian countries there is a much more extensive welfare system to provide generous out-of-work benefits to ensure that income inequality is not as high as in the US and UK. But it isn't just people that suffer from high energy prices. Businesses that compete in international markets and use a lot of energy might not be able to compete with firms based in areas with low energy taxes. This idea, called 'carbon leakage', explains why businesses ask for, and often receive, dispensation from energy taxes. This is particularly a risk when a small country or a state in the US unilaterally imposes high energy taxes. In the EU-ETS's third phase, permits are being auctioned off rather than being given away. Many energy-intensive firms complained that this would hurt their international competitiveness. This is a good argument for industry to make, but is it true? This will depend on how much energy costs really matter to the business and also how easy it is to bring the goods in from countries outside the tax zone. Tap water, for instance, is never traded between continents so a tax on water isn't likely to hurt the competitiveness of European water companies.

Figure 7.4 shows the cost of energy as a proportion of sales for some of the most energy-intensive sectors. The cost of energy is more than a tenth of the revenue from cement sales. Higher energy taxes are likely to be passed on to the customer if the price of energy rises. But so what? Cement is bulky and expensive to transport long distances: construction will have to pay more, or better still use less cement. We are not about to start importing cement from China – the

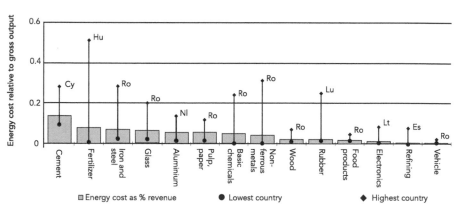

Figure 7.4 *Energy costs as a share of sales for the most energy-intensive industries in Europe, 2005*

Note: Two letter abbreviations signify the EU country with highest emissions. Cy – Cyprus, Hu – Hungary, Ro – Romania, Nl – Netherlands, Lu – Luxembourg, Lt – Lithuania, Es – Spain.

Source: E3G Briefing (2008) 'Ten reasons why giving free ETS allowances will not protect EU jobs or competitiveness', data originally from Eurostat *Structural Business Survey* results for 2005

costs of transport are prohibitive. Nitrogen fertilizers and some basic chemicals are highly energy intensive and less dense and so easier to transport – so these might be vulnerable. But the overall picture is that most goods are not at risk from increased energy prices – especially if the tax is levied across many countries simultaneously.

Governments should be much braver about raising energy prices for industry, and less coy about listening to lobbying. Ideally we would like to see energy-intensive industries work through global trade associations (as the world's steel and aluminium industries are doing already). This is the opposite of community-level action but there are some things that have to be coordinated at the national and indeed supranational levels. Improving the energy efficiency of energy-inten-sive manufacturing is one such area.

Innovative Tariff Structures

Sustainable community energy needs investment by energy companies to build new networks, information technology systems and plant. Energy customers should pay the majority of the cost of the transition through higher charges on energy bills. Some of the costs should be borne by the taxpayer. There's noth-ing controversial in this. But how should the costs be shared between different customers? And can we use the tariff structure in a smart way to encourage be-havioural change and investment in smart technologies?

A *dumb* tariff structure simply asks every household to pay the same sum of money and has the same effect as the UK's hated 'poll tax'; it's basically a tax on having an energy meter. Rich and poor pay the same rate; energy used on a windy summer evening and on a balmy winter evening would cost the same ir-respective of the huge difference in the production costs. We want the design of the tariff structure to support the broader social objectives.

Homes and small businesses pay a fixed price for every kilowatt-hour of en-ergy recorded by their meters. This covers the costs that vary with use, such as generation costs. The other costs of providing energy are usually recovered through the fixed charge, or in the first few units of consumption. The purpose of this charging structure is that the energy companies want to recover the costs of the wires and service from the fixed charge, and recover the cost of producing the energy from the variable tariff.

The picture is actually a lot muddier that this. Many energy companies also offer reductions to those paying by direct debit, or to those buying gas and elec-tricity as a bundle. But not all poor people have bank accounts. So those on low incomes end up paying more per unit of energy than do richer households. This is clearly unjust, and also environmentally perverse. The individuals who need to be encouraged to stop wasting energy are the comfortably off, not the poor. Energy companies also charge people using prepayment meters more (where tokens have to be fed into the meter, or the meter electronically topped up before energy is provided). Low-income households often choose to have prepayment meters as it enables them to avoid running up debt. Indebted customers are sometimes required to use prepayment meters by their supplier.

Innovative charging structures have been developed to incentivize more sustainable energy use or make energy more affordable to the poor. *Time-of-use tariffs* vary the price of electricity according to the time of the day to encourage people to use electricity when it is most in surplus (like the middle of the night). *Rising block tariffs* turn conventional pricing on its head; households' first units of consumption are cheap, and the next block of consumption is more expensive. *Social discounts* are mandatory subsidies that energy companies have to offer to vulnerable households.

Time-of-use tariffs

The idea behind time-of-use tariffs is simple enough. The demand for electricity varies across the day, in the early evening when offices and homes are both in use, it is much higher than in the middle of the night when people are asleep. It also varies between seasons: in hot countries the peak is in summer, in cold countries demand is highest in winter. But electricity is difficult to store. (This is unlike gas, where it is fairly cheap to store enough to smooth out changes in demand over the day.) The consequence of this variation in demand and difficulty of storage means some electricity plant is only used for a very few hours a year and is idle most of the time. Many large industrial customers voluntarily agree that the grid's system operator can temporarily switch them off to help balance supply and demand. They get handsomely rewarded for being retired like this. This is called responsive demand.

Time-of-use tariffs use pricing to persuade customers to move some of their demand to times when demand is light. Smart meters have been introduced to record and communicate energy used every half hour so different prices can be applied. Well that's the idea. An analysis in the UK by Owen and Ward (2010)[24] suggests that around a quarter of household energy could be shifted a few hours in this way. This is electricity used by washing machines, irons and fridges, but the modelling suggests that between only 2.5 and 5 per cent of demand will be shifted if consumers are left to get on with it voluntarily.

In countries with warmer climates this shift might be more pronounced as air conditioning units are enlisted in the quest to shift demand. A study in Australia[25] in 2006–2007 revealed that efforts to make air conditioning more expensive at peak times did reduce peak electricity consumption by 19–35 per cent where direct load control demand management was utilized.

But the real opportunity is for appliances to automatically power down when prices are high. Such smart devices would detect times of high tariffs and switch off or at least to low power mode. The most exciting such device is the charging of electric cars. Much of the forecast rise in electricity demand arises from such vehicles. If the car only operates for two or three hours a day, it can function as a rechargeable battery the rest of the time, supplying or drawing down electricity. This would be a fantastic opportunity – and will suddenly allow customers to act as the grid's balancing service, complementing wind power and other intermittent energy sources. This is still some time off.

Here and now small CHP units can operate as 'demand-side' smart devices. If these are used to provide hot water (which can be cheaply stored in a tank) and also export power, it doesn't much matter what time the plant is operating. If the system operator uses them to provide the grid with balancing services, this means there is less need for underutilized peak plant. Whether this is feasible depends on the complexity and rules of the electricity market. A small 20MW town CHP might be below the radar for a system operator who needs to acquire 1GW of power in a hurry. But as we've learned with distributive computing, it's possible for a network of small personal computers in aggregate to have the same computational muscle as a mainframe and at a fraction of the price. Smart grids could facilitate this.

In Germany one small CHP unit is doing exactly this.[26] In the EU Commission's MASSIG (Market Access for Smaller Size Intelligent Electricity Generation) programme, (the title is a good clue of what the Commission is seeking to achieve), one project tracks a small 4MW CHP plant selling power solely to the European Energy Exchange spot market. This is the energy market equivalent of sending grandma to play poker with Al Capone. Over the year the plant managed to sell power for an average price of €5.4/kWh, compared to the average spot price over the year of €4.0/kWh and the plant made more than twice the operating profit of the heat-only boiler. There were carbon savings too, and these would be bigger if biomass was used instead of gas. But just consider what this example shows us – a small village-level CHP plant can produce heat, power and grid balancing services in theory at zero carbon. The report's author Anders Andersen told us he is operating a flexible CHP plant in northwest Denmark that work in concert with a wind turbine, smoothing out its intermittency. This is crucial in the area as wind dominates local power production and the grid is not strong enough to bring in sufficient power from outside the area.

Just now smart meters and smart tariffs are meeting with a mixed response. Ontario has just introduced mandatory time-of-use tariffs. These are very smart, smarter maybe then many consumers: different rates are charged for peak 8.8c/kWh, mid-peak 7.2c/kWh and non-peak 4.0c/kWh. Further, the time of peak varies over the seasons, and between weekdays and weekend and holidays. The Ontario Energy Board hopes that time-of-use charges will avoid them having to build expensive plants used only at peak time.

Customers are going to have to get used to quite a lot of change. But this can be done sympathetically. SMUD is introducing smart meters only gradually and investing a lot of time in making sure customers understand how to operate their homes differently, and ensuring that it gets its customer service right. An installer fits the meter at the time agreed with the consumer and they promise to be no more than an hour late. They're making time-of-use charges optional rather than mandatory, wishing to avoid the backlash that another Californian utility faced in spring 2010. SMUD sees the smart meter as a good thing, even without time-of-use charges, since it reduces the cost of meter reading and speeds up the detection of line faults.

We think time-of-use tariffs are a useful means of allowing individuals and communities to participate in improving security of supply. But the costs to the consumer are up front, while the benefits are longer term and depend on electric vehicles and smart devices coming onto the market.

Rising block tariffs

Rising block tariffs make economists come out in a rash. Markets give you discounts the more you buy, social engineers do the opposite. Rising block tariffs are definitely the preserve of social engineers. The reason for this can be adduced from Figure 7.1. This shows that on the whole poor households use less energy.

The Dutch government introduced the concept of rising block tariffs when they introduced a carbon/energy tax for household customers in 1996. This included a tax-free energy initial allowance to bring down the cost of the first units of gas and electricity. In 2001, this was replaced by a fixed tax reduction of €142 a year for electricity. In the Flanders region of Belgium, every household receives 100kWh of electricity per household, plus a further 100kWh per family member for free. Pensioners receive the first 500kWh free. These free allocations mean that consumption above the ration costs more.

Many US states have used two-tier tariffs so that consumers pay about 15 per cent more for units in the higher tier than in the lower. After the California electricity crisis in 2000/2001, the three regulated utilities – PG&E, Southern California Edison (SCE) and San Diego Gas & Electric (SDG&E) – needed to raise substantial revenue, but the regulators were worried about the impact on lower-income households and so just allowed the price rises in the upper tiers. One report by Borenstein (2008)[27] notes the huge increase in price for the upper tiers: 'Regulators adopted a five-tier increasing-block retail pricing structure ... By 2008, the price of the highest block ... ranged from about 80 per cent higher to more than triple the price on the lowest block, depending on the utility.' Borenstein analyses the distribution impact of these steep increases:

> the steep five-tier rate structure benefits low income customers, but that benefit is fairly modest. For PG&E and SCE, which have the most steeply increasing prices – the best estimates suggest that returning to a two-tier structure similar to the pre-2000 utility tariffs in California would raise the annual electricity bill of the poorest households by slightly less than $100 per year or about $7–$8 per month.

Rising block tariffs have yet to be used for gas. The UK's Climate Change Committee published a paper in 2009 arguing that rising block tariffs for electricity would be good for the fuel poor (except for those not on the gas grid who use electricity for heating), but that gas rising block tariffs would be bad for them.[28]

Rising block tariffs are an important way to promote energy efficiency and to encourage rich households to notice how much fuel they are using. They should therefore be introduced for both gas and electricity. Both should be designed by regulators and governments to protect those on low incomes or particularly vulnerable to electricity or gas price rises. For example, those not on the gas grid who use heat pumps should be able to opt out of electric rising block tariffs. Those who are ill or elderly and need to keep their homes warm (and may have to spend longer at home) should also be able to opt out.

Social tariffs

Governments sometimes request or mandate (or, as in the case of the UK, begin with a request and switch to a mandate when that fails) that vulnerable households be offered a discounted tariff. There should also be mandatory social tariffs offered by all energy suppliers. California mandates the CARE (California Alternate Rates for Energy) programme to provide discounts of around 20 per cent to low-income customers. A quarter of all the customers of the three regulated utilities benefited from the CARE programme in 2006.

None of these tariff structures have occurred spontaneously through the competitive electricity market. All are contrary to market forces and have had to be mandated, or, in the case of time-of-use tariff/smart metering, require a high degree of coordination to ensure an orderly roll-out of the technology. The FiTs described above also need heavy regulation; electricity companies would otherwise shun customers who generate their own power.

This does raise the question: what is the point of retail competition if it stifles what we consider to be socially desirable outcomes and we have a tapestry of behind-the-scene payments to even out the cost of meeting social and environmental price incentives? All of these require heavy manipulation of energy prices.

We think supply and distribution should be integrated so that demand-side incentives can be geographically coordinated and local communities have a single entity to collaborate with. The liberalization of energy markets has been a flop – its continuance will retard the introduction of a low-cost smart grid. Even on its own terms of providing consumers with competition and choice it has failed. Most consumer organizations can rattle off horror stories of bad practices, including energy company salespeople tricking customers into switching to poorer deals. In the UK the monopoly water companies are preferred to the competitive energy markets.

Conclusions

The price of energy will ultimately rise as we start to run out of oil and gas over the next ten years or so. We have found ways of accessing gas more easily but it's still going to run out. We got a taste of how quickly energy prices rise between 2006 and 2008 when demand from China and India grew faster than supply could respond before the recession in the West.

But rather than waiting for markets to raise the price of energy uncontrollably, we argue that government should tax energy to curb its consumption and raise much-needed revenues. We support a carbon tax (or an effective, auctioned ETS with a tight cap) for a number of reasons. First, some of these revenues should be used to assist the vulnerable households and support technologies that need support. Second, it is important that investors are convinced that the policy is enduring and that governments in ten or more years will not reverse it. A carbon tax that raises tens of billions cannot easily be replaced by another income stream.

Notes

1   Miliband, E. (2010), www.publications.parliament.uk/pa/cm/cmhansrd.htm.
2   Stern, N. (2007) *The Economics of Climate Change: The Stern Review*, Cambridge University Press.
3   *The Guardian* (2007) 'Climate change a "market failure"', *The Guardian*, 29 November.
4   See www.turn.org.
5   Interview with Ian McMillan (poet), *Desert Island Discs*, BBC Radio 4, 7 November 2010.
6   See www.vizelia.com/en/actualites_breves.php.
7   It is worth reading our friend Nick Robins (2007) *The Corporation that Changed the World – How the East India Company Shaped the Modern Multinational*, Pluto Press, for an excellent exposition of this atrocious company.
8   Department of Energy and Climate Change (2010) 'Fuel poverty monitoring indicators 2010', Department of Energy and Climate Change, London.
9   Scottish Executive (no date) *Scottish House Condition Survey*, www.scotland.gov.uk/Topics/Statistics/SHCS.
10 Zero Energy Ltd (2008) *Fuel Poverty in Great Britain, Germany, Denmark and Spain: Relation to Grid Charging and Renewable Energy*, report prepared for Highlands and Islands Enterprise.
11 Centre for Sustainable Energy and Association for the Conservation of Energy (2010) 'Distributional impacts of UK climate change policies', final report to Eaga Charitable Trust. The original data underpinning the diagram were drawn from aggregated data from the 2004 to 2007 *Expenditure and Food Survey* carried out by the Office for National Statistics.
12 Cited in West Midlands Public Health Observatory (no date) 'Fuel poverty and older people', based on 1997 study by the third Eurowinter group.
13 Eurostat (2010) 'Electricity prices for second semester 2009', Eurostat 22/2010, Luxembourg, www.epp.eurostat.ec.europa.eu/portal/page/portal/product_details/publication?p_product_code=KS-QA-10-021.

14 Eurostat (2010) 'Electricity prices for second semester 2009', Eurostat 22/2010, Luxembourg, www.epp.eurostat.ec.europa.eu/portal/page/portal/product_details/publication?p_product_code=KS-QA-10-021.

15 Knigge, M. and Görlach, B. (2005) *Effects of Germany's Ecological Tax Reforms on the Environment, Employment and Technological Innovation*, Ecologic Institute for International and European Environmental Policy gGmbH.

16 Speck, S. (2009) 'Ecological tax reform in Europe: Experience to date', Green Fiscal Commission, www.greenfiscalcommission.org.uk/pdf/london2009/Stefan Speck.pdf.

17 Authors' calculation based on Ofgem's Factsheet 78 (2009) 'Wholesale and retail energy prices explained', Ofgem, www.ofgem.gov.uk/MEDIA/FACTSHEETS/Documents1/wholesaleretailpricefactsheet.pdf .

18 Ragwitz, M. and Huber, C. (2007) 'Feed-in systems in Germany and Spain and a comparison', Fraunhofer Institut für Systemtechnik und Innovationsforschung, Karlsruhe, www.bmu.de/files/english/application/pdf/langfassung_einspeisesysteme_en.pdf.

19 Federal Ministry for Environment, Nature Conservation and Nuclear Safety (2010) 'Key information at a glance 2009', www.bmu.de/english/renewable_energy/doc/39831.php.

20 Gipe, P. (2008) 'Summary statistics of emissions avoided, fossil fuels offset, imports avoided, and overall cost savings in Germany during 2008 from feed-in tariffs', www.wind-works.org/FeedLaws/Germany/SummaryStatisticsofEmissionsAvoidedandFossilFuelsOffsetinGermany2008.html.

21 California Senate Bill 1078 2002, since amended by Senate Bill 107 2006 and Executive Order S-14-08 signed in 2008.

22 Department of Energy and Climate Change (2010) *Warm Homes, Greener Homes: A Strategy for Household Energy Management*, Department of Energy and Climate Change, www.decc.gov.uk/en/content/cms/what_we_do/consumers/saving_energy/hem/hem.aspx.

23 Healy, J. D. (2003) 'Excess winter mortality in Europe: A cross country analysis identifying key risk factors', *Journal of Epidemiology and Community Health*, vol 57, no 10.

24 Owen, G. and Ward, J. (2010) 'Smart tariffs and household demand response for Great Britain's Sustainability First', www.sustainabilityfirst.org.uk/docs/2010/Sustainability%20First%20-%20Smart%20Tariffs%20and%20Household%20Demand%20Response%20for%20Great%20Britain%20-%20Final%20-%20March%202010.pdf.

25 See www.etsautilities.com.au/centric/our_network/demand_management.jsp.

26 Streckiene, G. and Anderson, A. (2010) 'Analysing the optimal size of a CHP-unit and thermal store when a German CHP plant is selling at the spot market', MASSIG EIE/07/164/S12.467618, unpublished.

27 Borenstein, S (2008) 'Equity (and some efficiency) effects of increasing-block electricity pricing', http://econ.as.nyu.edu/docs/IO/8821/Borensteinpaper.pdf.

28 Committee on Climate Change (2009) *Rising block Tariffs and the Fuel Poor*, Committee on Climate Change, London.

Chapter 8

Nudging and Shoving People and Business into Changing their Behaviour

A very Faustian choice is upon us: whether to accept our corrosive and risky behaviour as unavoidable ... or to take stock of ourselves and search for a new environmental ethic.

E. O. Wilson[1]

Society will change and adopt more sustainable habits only after a reasonable size environmental disaster ... a disaster that kills a few hundred thousand people like us ... people from a European or North American country.

Remark by a senior environmentalist at a conference

The two quotes hint at many environmentalists' view as to our capacity to change our behaviour for the greater good. The second remark darkly concludes that people will only change once galvanized by a real and present threat to their lives. One fed-up senior official remarked to us: 'When someone says the answer to any public policy question is: we need culture change, or we need people's values to change, I feel like reaching for a gun and shooting them'.

We struggled for a long time about whether to include a chapter on behaviour in the book. Isn't all public policy about changing people's behaviour? Does behaviour change really merit its own treatment? We decided to include it. Humans are social creatures and imperfect consumers. Behavioural change harnesses these two traits for public purpose.

In Chapter 7 we argue for direct financial incentives to align commercial and consumer decisions with saving energy. Incentives have flaws. Meaningful incentives mean transferring large dollops of money from one group of people to another – complex delivery architectures have to be created to look after poor energy customers or stop internationally trading firms from being damaged.

Another approach is using regulations. Regulations can be intrusive and expensive to police. Both incentives and regulations are necessary, but that doesn't mean that we can forget about behaviour change. Behaviour change is the only feasible way of implementing many minor lifestyle adjustments that save energy. There is no way of regulating people to set aside their clothes dryer on a balmy summer's day, no feasible way of compelling people to only heat rooms that they use, or to carry out the numerous fiddly jobs around the house that save energy, draught-stripping windows and doors, putting jackets on hot water cylinders or thermal linings on curtains, and remembering to draw them at night.

Behavioural change means different things to different people. In this chapter we talk about three sorts of change, encouraging greater direct *take-up* of energy-saving behaviours in homes and businesses; participation in *community decisions* about local energy use; and people *applying and learning* low-energy techniques in their professional lives.

Greater Take-up of Environmental Behaviours

There's a huge gulf between what people profess to think and what they do in practice. This difference in belief and action is well known and found throughout the world. Figure 8.1 shows the results of research by Leiserowitz et al (2010) from Yale and George Mason University.[2] The majority of Americans understand broadly which actions are important in terms of environmental benefits: turning off lights and adjusting thermostats are correctly viewed as important, carrying beverage containers and unplugging electronics are correctly seen as unimportant (though people seem to be in denial about their transportation behaviour). But the actions that are actually performed are more to do with convenience and avoiding disruption to everyday habits. Reducing the amount of waste being produced, accepting colder internal temperatures in winter and recycling at home requires modest changes in everyday life, but sadly only about a half the number of people who believe them to be important actually act accordingly.

Europeans think that keeping energy prices affordable, especially for elderly people or families with young children, and security of supply, followed by environmental issues should be the most important priorities for energy policy. The desire for cheap energy and sustainable energy, which we think to be oxymoronic, is rationalized by stressing the importance of energy conservation: 54 per cent regard reducing energy consumption as very important[3] and only 2 per cent see it as unimportant.

There is also inconsistency in what Europeans say and what they do. Box 8.1 discusses an example of the difficulty people have in acting on their beliefs. Around half of Europeans say they plan to spontaneously reduce their energy and so are not prepared to pay more for their energy use.[4] But when asked which actual actions they undertook over the past 12 months, 48 per cent said they cut down on lighting and use of electrical appliances and 42 per cent on heating or

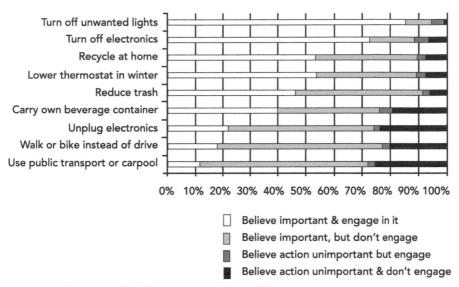

Figure 8.1 *The attitude–behaviour gap in US adults*

Note: 1024 adults were interviewed.
Source: Leiserowitz et al (2010)[5]

air conditioning. But few other actions were taken: only a fifth cut back on vehicle use or insulated their homes; the same number said they had done nothing at all.

People broadly favour renewable energy technologies over fossil fuels. The favoured sources of energy are solar (80 per cent support), wind (71 per cent) and hydro (65 per cent); at the other end of the popularity stakes are nuclear (20 per cent support), coal (26 per cent), oil (27 per cent) and gas (42 per cent). The ranking of the different technologies' popularity is fairly stable between countries. Wind is the second most popular source of energy in the UK after solar; it is twice as popular as gas, which is the least disliked fossil fuel. Most Europeans are prepared to pay more for renewable energy – but only a quarter of people are prepared to pay more than 5 per cent more for renewable energy.

How do we get people to act in environmentally friendly ways? There is a tremendous amount of information available to people on how they can reduce their environmental impacts. And government agencies spend lots of money on education campaigns about environmental problems and tips on how to address them. However, information alone doesn't work. People feel overwhelmed by the amount of information, feel that what they could do would make little difference, are not aware of how easy it is to make daily habit changes, or are too busy to attend to anything that demands time and effort.

Changing people's behaviour is possible. But people differ from one another: in their capacity to change, interest in energy issues, political views, personalities, income, available free time, skills and knowledge, tenure of their housing and connectedness with others in their community. In order to engender behaviour

---

Box 8.1 The gap between knowing and doing

It was 12 January 2010. The whole of Western Europe was experiencing one of the coldest winters in 40 years. Records were tumbling: unprecedented snowfall, coldest recorded temperature in Dublin, shortages of gas through the interconnector between UK and Norway. There was talk about gas supply being withdrawn from firms that were contracted to interruptible tariffs – which means they bought gas for a discount on the understanding that when supply conditions were tight their supply would be switched off.

Staff in UK's Consumer Focus had spent the day thinking about the effects of the freezing weather on customers. Would the elderly and vulnerable be able to keep warm? What would gas prices do in the next few days? That evening a number of staff, including Prashant, were still in the office in one of the meeting rooms. One said that she was stiflingly hot and asked if anyone minded if she opened the window. No one demurred. No one knew how else to bring the temperature of the room down to a comfortable level.

It would be great if this was an unusual case, but all of us have been into rooms where lights are on and daylight is gushing through the windows, or we left the heating or the air conditioning on and left the home vacant for a day. Each individual oversight is not a lot, but they add up. Careful management of internal temperature and lighting accounts for much of the difference in energy consumption between homes and the differences can be substantial.

For a long time we have treated this as though it were a failure of education or awareness. But even if that were so, many of the efforts of the public authorities seem laughable. Public service advertisements in the UK used to chide us about energy efficiency – the authors remember an enigmatic catechism used by the energy efficiency agency at the time 'Do your bit' printed on sugar sachets – a bit too subliminal for us and indeed for most others. But the diagnosis of lack of knowledge is too convenient, too straightforward to be the whole story. No one in Consumer Focus that evening was unaware of the pressures the country's gas supply was under, nor the more general issue of climate change.

---

change, the UK's Department for Environment, Food and Rural Affairs identified seven segments in its Responsible Environmental Behaviour programme.[6] It did so through commissioning and synthesizing a number of qualitative and quantitative reports. The seven segments are: positive greens, 18 per cent; waste watchers, 12 per cent; concerned consumers, 14 per cent; sideline supporters, 14 per cent; cautious participants, 14 per cent; stalled starters, 10 per cent; and honestly disengaged, 18 per cent. Positive greens are early adopters and pioneer many environmental behaviours; they're not overly concerned with saving money. Waste watchers are motivated by a desire for thrift and dislike of needless waste. Sideline supporters are mildly concerned with environmental issues, but not enough for them to do much change their day-to-day lives. The last three categories are largely unconcerned about environmental issues and the last two are actively hostile, either because they have much more important priorities or they disagree with the evidence and don't feel it their responsibility to act. The Department has suggested that public policy can catalyse behavioural change by adopting a mixture of encouragement, enablement, engagement and exemplification. How much of each of these four ingredients to apply depends on the behavioural change being sought and the segment of the population being targeted.

The Canadian community-based social marketer Doug McKenzie-Mohr's approach to fostering behavioural change is similar to that of the Department for Environment, Food and Rural Affairs. First, segment the community, then identify precisely the behavioural changes being sought and then make use of a few behavioural tools to encourage the change. The type of tools includes appeals to *commitment* (and then *self-consistency*) from the individual, deploying *prompts* or *reminders* close in time or space to the action we are trying to influence, creating new *norms* to exploit people's wish to conform, *effective messaging* so information is vividly portrayed, selecting a *credible source* so the message is delivered by a trusted source, *making it easy to act*, which requires considering and removing barriers, and *rewarding* progress with incentives, which do not have to be financial and can include noticing and acclaiming good deeds.

There are many other behavioural models that identify points of leverage for public policy-making. There is a current vogue for applying behavioural economics in public policy. This uses the idea that people are imperfect consumers, nothing like the *Homo economicus* assumed in economic textbooks. People make systematic mistakes in their choices because they are imperfect consumers plagued by time myopia, unhealthily averse to loss, unable to assess risk properly, poor at choosing from are a large number of alternatives, too short of time to properly research new ideas, and overly influenced by irrelevant details such as celebrity endorsements. Far from knowing precisely what they want and being able to effortlessly navigate all the thousands of choices, people have pliable and sometimes irrational demands and hence make far-from-optimum purchase decisions.

One of the major policy innovations arising from the discipline of behavioural economics is the idea of government specifying that goods should incorporate default options that meet the average person's needs but which people can choose to vary, should they so wish, to better reflect their preferences. The best example of this in the UK is the NEST pension, which is being introduced from 2012. Employers automatically enrol their staff into the scheme, and the payments are saved in a portfolio of different types of asset classes suited to the employees' age. Investments for each worker are decided paternalistically by the use of profiles of employees. Public policy-makers are trying to take advantage of these behavioural biases to 'nudge' people into making decisions better aligned to their long-term interests.[7] In the UK the Cabinet Office has published a report called 'MINDSPACE',[8] which provides a checklist of behavioural incentives that policy-makers are being encouraged to use to influence behaviour.

There are a huge number of social initiatives and programmes that have sought to change people's behaviour. We select some fun and high-profile examples from UK and North America as examples.

When people make a pledge they are more likely to change their behaviour. A famous pledging tool is the UK's 10:10 campaign. This is an international social movement that started in 2009. The numbers '10:10' represent the goal of the movement – to cut carbon emissions by 10 per cent in 2010. Unless its activities are fundamentally incompatible with the emissions reductions goal, any individual or organization can join by signing an online pledge to cut their emissions.

In presentations, the 10:10 campaign staff say the 10 per cent figure was chosen partly for its ease of communication and partly because it makes the goal achievable. The year-long timeframe encouraged people to begin acting immediately. Additional objectives of the movement are to encourage government policies that address climate change and to shift public attitudes towards climate change by normalizing low-carbon activity. Enrolment is simply by asking people to electronically pledge their commitment. At the time of writing there were nearly 110,00 people, 3822 businesses, 2155 schools, 210 local councils and 2100 other organizations engaged in the 10:10 movement. Campaigns now are running in over 152 countries.

Participants can report their energy use using an online carbon calculator. To help people succeed in reaching the target, 10:10 staff provide case studies and practical advice on their website. Supporters also organize many events to promote the movement such as concerts, races, festivals, etc. On 10 October 2010 in The Netherlands, the country's largest TV channel hosted a 90-minute '10:10 takeover' in the form of a quiz show in which ten Dutch celebrities were quizzed on their environmental knowledge.

The UK's official energy-saving agency, the Energy Saving Trust, operated a similar pledge-based system, the 'I commit' campaign, between 2006 and 2008. It asked people to make a few simple promises that would reduce their energy use by 20 per cent. Three-quarters of the commitments were simple: switching to energy-efficient light bulbs, only filling the kettle with the required amount of water, washing the laundry at 30°C. A quarter of the commitments were for more substantial changes such as installing loft or cavity wall insulation. Around 200,000 (around 0.3 per cent of UK's population) made a commitment. All the names of participants were available on the website. It encouraged actual action by asking for email addresses of friends who would be told of the participant's commitment and encouraged to firm up resolve.

Pledges can work. But the main problem with both these schemes was that no real effort was made to follow through, encourage participants or check they had done anything at all. Making a promise on a website is as binding as the promise of undying love to a stranger after too many drinks. Both programmes made use of the cheapness and ease of the internet to access large numbers of people, but the commitments made were shallow. The follow-through was often just an impersonal email reminding the participant of their pledge (and possibly asking for money), more likely to be deleted than to inspire change.

Commitments can be built on. A good salesman will follow through on the sale of one item and use it as a platform for developing a relationship with the client. This is the 'foot-in-the-door' strategy. People who have agreed to a small request are then more easily influenced to comply with larger requests. Why? When people agree to a small request, indicating, for example, support for energy efficiency, it changes their self-concepts. They come to see themselves as the kind of people who support the cause. When asked later to comply with a larger request, there is internal pressure to behave in a way that's consistent with their self-concept. Local community groups are able to do the follow-through and inspire people to persevere with their promises. In a local group that Prashant is involved with, all of us have talked about insulation measures we have promised to undertake in our houses. We take it in turns to host meetings and good naturedly show off our progress to one another.

The principle is exemplified well in the Twin Cities programme (see Box 8.2). Homeowners first are invited to a workshop. Then they are asked to allow an energy squad to come into their home. Finally, they are asked to make significant energy improvements in their homes and 20 per cent of them do.

Box 8.2 Twin Cities programme

Energy efficiency policy historically has focused on individual low-cost measures. In order to achieve the greater savings in carbon emissions that are increasingly important, we need to move towards a whole-house and energy-services approach. This is the goal of a programme initiated by the cities of Minneapolis and St Paul, Minnesota. In 2009, they formed a coalition with two non-profit organizations to implement a residential energy efficiency programme. Twelve neighbourhoods were selected. Programme funding came from two utilities and the state lottery. The programme involves several steps:

1   A specific neighbourhood is targeted for outreach.
2   Homeowners are invited to a free workshop. At the workshop, attendees are given energy efficiency information and free gifts such as low-energy light bulbs and faucet aerators. They are also told that they're eligible for a subsidized $400 home visit that will cost them only $30. In Minneapolis, over 95 per cent of homeowners attending the workshops paid the $30 and registered for a home visit.
3   A home energy 'squad' is sent to walk through the home with the homeowners and make small improvements immediately (for example wrapping water heaters with fibreglass blankets).
4   At the end of the home visit, an attempt is made to convince the homeowners to undertake comprehensive improvements. The squad offers assistance in locating contractors and tells participants about financing options. At present, 20 per cent of homeowners visited choose to make significant energy improvements.
5   Homeowners are given assistance completing major upgrades, including consultation with loan officers. They also are sent a bi-monthly energy progress report that tracks their energy use in comparison with similar homes.

Real-time electricity displays provide households with information on how much energy either individual gadgets or the entire house is using then and there. They can be used to detect if devices have accidentally been left on, or for working out which gadgets in the house or office consume the most energy. If displays are prominently located and they display information vividly (for instance with a big red lights when energy use is high), they can change behaviour. In one detailed study by Friedrich (2010)[9] undertaken in 2004, Hydro One in Ontario provided 382 customers with real-time display meters; no other supporting intervention was made. The test relied purely on customer interest and curiosity. Over the year customers reduced their consumption by 6.5 per cent compared to the control group. Savings were lowest in homes with electrical heating, suggesting the meters prompted the turning off of smaller devices such as lights. The savings were also shown to be persistent and didn't drop off over the year as people got used to them.

Use of prompts is not a panacea. There are many examples of messages capturing people's attention for a short while, but then people losing interest. Some people are simply not motivated, no matter how vividly information is portrayed. The use of distant and impersonal tools to provoke behavioural change only works slightly.

People are strongly influenced by the actions and views of others; they don't like to let others down and they like to conform to the expectations of those they respect. To make behavioural changes enduring, new energy-saving habits will have to form and people will have to start thinking this is the way things are done.

Donald Kelley launched a community-based energy reduction programme *Energy Smackdown* (see Box 8.3) and dressed it up as a game show. Funding comes from utilities and charitable foundations. In the 2010 'series', three teams of ten households in Massachusetts competed over a year to reduce their energy consumption. At the outset, each household received a free energy-use audit and lifestyle assessment. Each month they had to undertake new tasks: the light bulb challenge, air sealing round one, smart transit and so on.

Smackdown teams engage in two types of challenge. First, there is a household competition in which households reduce their carbon emissions as much as possible on an individual basis. Reductions are measured in electricity, heating fuel, waste, air travel, auto travel and servings of meat. Second, there is a team competition. The object here is to earn points for your team that are allocated for specific energy-saving actions (for example insulating an attic). Teams also work together during special publicized challenge events to earn points (for example holding a 'localvore banquet' in which teams prepare meals using local ingredients).

Teams can record their points and track their opponents' points online. There are also social events to spur enthusiasm, including kick-off, half-time and finale celebrations. Utility staff serve as energy guides, educating participants about home energy upgrades and financial incentives available. Participants also learn about energy choices through the processes of the initial energy audit, by winning points for performing certain actions, and by discussions in team meetings. Energy use is monitored on a monthly basis from utility and car fuel bills. The tribulations are broadcast on a local TV channel and short film clips can be found on the website.[10]

The *Energy Smackdown* approach attempts to instil new norms at the 'community' level, both among the participants and also the viewers of the show, through the local media.

Using people's competitiveness and desire to support team efforts has also been exploited by one of the UK utilities. British Gas piloted 'Green Streets' in 2007–2008 (see Box 8.4) as a competition between 14 communities around the UK. The pilot was rerun in 2010 on a larger scale. Each Green Street includes at least 20 households and a community facility. Most of the successful 14 candidates were pre-existing local charities or virtual groups. British Gas provides technical and financial support to help them deliver their ideas. Plans include energy audits, community decentralized energy sources such as heat pumps, and energy efficiency measures. In an interim analysis by the Institute for Public Policy Research,[11] several of the contestants said that they 'came along for the ride' rather than being greatly motivated by saving energy. One comment was 'By doing it as

---

Box 8.3 Energy Smackdown is?

… a reality TV show run by a not-for-profit educational charity, the BrainShift Foundation. It uses the game show format to pit three competing teams of households from different cities in Massachusetts (in series two, the competition was between teams from Arlington, Medford and Cambridge) who reduce their environmental impact through a series of actions they can take in their homes and lifestyles. Focus is on transport, building energy efficiency, electricity and heating. Each team works hard to reduce their emissions not just for their own sake but also for the sake of their team-mates in the other nine households. The teams provide valuable support and motivation to one another.

Each action taken is called a Nugget. In the second season these included reducing the hot water temperature, installing low-flow aerators in showers, low-energy lighting, line drying laundry, air sealing and locavore banquet (local food cooked at home). Heat and electric savings are verified through the monthly utility bills; N-Star the local utility facilitates this data sharing.

The TV show is screened on a public TV channel over a 12-month period. Series two had seven episodes and is available on DVD and on local access TV stations. The show tracks 30 households (ten from each city).

The average utility electricity bill was reduced by 14 per cent in series one. Series one contestants and their families report that they are maintaining their actions. The show has also motivated contestants to undertake actions not covered by the show such as insulating their roof space. In series two, CO_2 emissions were down by 33 per cent half way through the season.

Box 8.4 The Green Streets 2008 Competition

Green Streets was an energy use pilot in 2007–2008, led by British Gas, the UK's largest energy and home services provider. Sixty-four households – eight from one street in eight UK cities – were selected to participate in a year-long challenge to reduce energy use by as much as possible. The prize for the winning street was £50,000 to spend on a community project of the community's choice. British Gas asked the Institute for Public Policy Research to analyse the Green Streets data and to interview participants to draw out policy lessons.

Participants were brought together in several neighbourhood meetings and energy audits were conducted for each property. In addition, £30,000 worth of energy savings and renewable energy measures were given to each street, i.e. £3750 per household for insulation, efficient appliances and heating systems, etc. Each street had a British Gas energy efficiency expert to help determine the best use of the funds.

Analysis of the data indicated that the energy savings in the households were due both to the energy efficiency measures that were installed and to changes in energy-use behaviour. Interviews with participants suggested that the following were drivers of behavioural change:

- encouragement from the expert advisers;
- feedback from handheld, real-time display units given to all participants that showed how much electricity a particular appliance was using and how much it was costing;
- the competition among streets and the mutual support and peer pressure between other participants in the same street.

On average, the energy saving across all Green Streets households was just over 25 per cent. The averages for streets ranged from under 15 per cent in London to almost 35 per cent in Leeds. The latter was the winner of the competition.

a community rather than individually, people are talking about it and comparing notes, so there are behavioural changes that happen because of what your friends and neighbours are doing'.

Competition is an effective way to promote behaviour change. Most people have little idea about how their energy use compares to that of others. They lack a point of reference. A number of utilities are including information on their customers' energy use the same month a year ago and energy use by neighbouring households. This provides a useful reference point and can spur people to action.

A carefully controlled analysis of 80,000 households was carried out in Minnesota.[12] Half were sent home energy reports around three weeks after their energy bill was issued. The control group were not sent these reports. The home energy reports included social comparison data comparing their performance with the average and the top 5 per cent of their comparator group. If they exceeded the reference points their performance was described as 'great' and

Home Energy Reports: Social Comparison Module

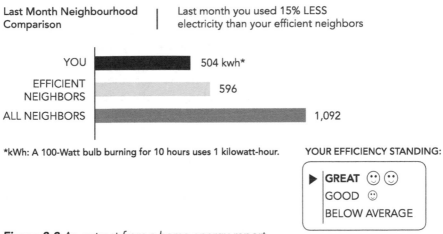

| Last Month Neighbourhood Comparison | Last month you used 15% LESS electricity than your efficient neighbors |

*kWh: A 100-Watt bulb burning for 10 hours uses 1 kilowatt-hour.

Figure 8.2 *An extract from a home energy report*

Source: Allcott, H. (2009)

'good'. The report also included 'action steps' suggesting tailored changes to their energy use guessed from the pattern of energy usage. These included some quick wins, small energy-saving purchases and longer-term ideas, for instance if a family displayed high summer energy use the company suggested they buy new air conditioning. Figure 8.2 comes from a home energy report. As a further twist, some people were sent reports monthly and others on a quarterly basis. On average people sent monthly energy reports cut their consumption by 2.1 per cent and those receiving quarterly reports cut their energy use by 1.6 per cent. The effect was most pronounced for homes that used more than the average amount of energy. If no information was received for more than a month, the behaviour reverted back to normal.

This approach was repeated in Sacramento by SMUD. SMUD's partner OPOWER sent 35,000 customers information on their energy usage compared to that of people in neighbouring blocks and their own historical usage; 50,000 control customers were also monitored. Savings were highest for those consumers using more energy than comparable households. Savings were unrelated to other factors such as income, ethnicity and household characteristics. OPOWER uses emoticons (☺ or ☹) on bills to keep 'high savers' motivated and avoid persistence drops.

Use of bills with social comparison information works best for those who use more energy than their comparators. These are the people it is often hardest to motivate. The 10 per cent of households with the highest energy usage reduced their consumption by 7 per cent. For the people who used less energy than average for their comparator group, the savings were less than 1 per cent.

This should come as no surprise. High-energy users use twice as much as their neighbours and the bills simply draw this to their attention. They start to notice that lights are left on unnecessarily or that the air conditioning or ducting is faulty if their summer bills are high. Simply drawing high usage to their attention has a huge impact. The lowest energy users already pay attention to many of these sorts of easy savings.

People who are already *positive greens*, to use the Department for Environment, Food and rural Affairs' terminology, can play a positive role. These people can influence others around them by setting an example. But they too can have attacks of loss of confidence, despair and feelings of isolation. Numerous initiatives have sprung up where positive greens band together and support one another's efforts. In the UK there are active Climate Action Networks, Carbon Conversation groups and Transition initiatives (also known as Transition Towns).

Transition Towns (see Box 8.5) is a now becoming an established community movement, in which people coalesce around a shared imperative to change their lives to become more sustainable. The movement is based on the assumptions that disruptions from climate change and peak oil are inevitable and that we urgently need to act to survive in post-fossil fuel communities. In essence, the goal is to re-localize elements that a community needs to sustain itself, for instance growing food locally and learning low-carbon (and survival) skills from older people. In north London the local Transition Town has received funding from the local government and provides basic environmental training to residents. Prashant attended a draught-busting workshop where his inability to hammer a nail straight was rudely exposed to his equally inept, largely white-collar and professional neighbours. The Transition Town trainer said that about ten residents attend the average workshop; around half of these follow through and collect the free insulation material. So at the cost of around £300, the council had successfully encouraged the draughtproofing of five homes.

The premise of the movement – that climate change and peak oil will have grave and enormous consequences for human civilization – presents an ominous spectre, which some will find off-putting. When we perceive a threat of any kind, we make two appraisals. First, how dangerous is it? Second, do I have the resources to handle it? When we judge that the threat is very dangerous and we believe we don't have the resources and ability to manage it, we try to control the emotion itself. We engage in something called 'emotion-focused coping'. That is, rather than attempting to manage the threat by taking action, we try to manage our emotions instead by engaging in techniques such as denial. This, in fact, is an important reason why many people don't take action on the issues of climate change and peak oil. Transition initiatives, however, provide a sense of efficacy – a sense of power in numbers. Individuals work collectively to determine specific actions they can take to address the threat. A guiding principle is positive visioning – the emphasis is on empowering possibilities and opportunities.

Box 8.5 Bottom-up Transition Towns

People commonly argue that our values shape our behaviour. The reverse is also true. When we see ourselves acting in a way that indicates we support a certain cause, our values then change; we come to believe that the cause is good, right and important.

Transition Towns originated in 2005 in Kinsale, Ireland where a group of college students prepared a report for a permaculture course taught by Rob Hopkins. Hopkins has written *The Transition Handbook*,[13] which sets out how communities can make the transition from a high-energy consumption town to a low-energy town. It laid out a vision of a future released from food supply chains that depend on fossil fuels. The results were presented to the Kinsale Town Council and the overall plan was approved for adoption.

The work of the initiatives varies a great deal. Examples include organizing neighbourhood gardens, building solar panels, hosting networking fairs, creating wild plant food maps and writing grants proposals for projects, such as retrofitting houses. Initiatives gladly work with local governments and welcome government officials. They emphasize, however, that the role of government is to support and not drive the initiative.

The Transition model involves 12 'ingredients' that are adapted to meet local conditions. These include raising awareness of climate change and peak oil among members and the broader community, networking with other local groups and facilitating reskilling. Often sub-groups will form to focus on particular issues including food, waste, energy, education, youth, economics, transport, water and local government. They also have an objective of creating and implementing an 'Energy Descent Action Plan' by convening the work groups.

There are over 300 official Transition initiatives, about half of which are in the UK. The US has developed its own network based on the European model that included 75 local initiatives as of October 2010.

Home Improvements

Some of the necessary 'behavioural changes' are not simply a matter of noticing the actions or practices that waste energy. Small changes can only do so much. Once these are exhausted people have to start improving the fabric of their homes and offices. This means summoning the energy and time to plan how to improve their home's energy efficiency.

Few people seek out information on how to improve the energy efficiency of their homes or workplaces. They have better things to do with their time. Those who do seek information often find it lacking, it needs to be tailored to their budgets, family plans and tastes. Who should they trust to give accurate advice about which technologies are appropriate for their homes and which grants are available from companies and government? The bespoke nature and trustworthiness of advice is a problem with many mass-market energy efficiency messages.

Many governments have websites and telephone help lines on energy efficiency. But energy-saving agency staff confide that the leaflets, internet or telephone-based advice often provide little measurable increase in the take-up of technologies, though they probably do reassure people who have already decided to make changes. A report by the UK's National Audit Office[14] found little evidence of people acting as a result of the advice.

Most people need to see and feel the changes that are possible before they have the confidence to spend money on their own homes. John Doggart runs an educational charity, the Sustainable Energy Academy, to showcase 'Old Home SuperHomes' that have undergone deep retrofits to improve their energy efficiency and installed micro-renewables. Homes are selected if they have achieved cuts in energy use of at least 60 per cent. He says: 'My ambition is that everyone should have an exemplar homes within walking distance of where they live'.[15] The charity runs open-home weekends so visitors can be shown around the homes. The owners or volunteers from the community, under John's instruction, guide people around the houses pointing out the features, explaining how much each costs and giving frank views on any challenges they faced, for instance, finding suitable tradesmen. Usually there will be leaflets or materials to show the products that were used. At the time of writing there are 70 homes dotted around the UK – most of them privately owned, but some owned by local government. Most are open to visitors several weekends a year.

Prashant acted as a volunteer guide for a SuperHome in London. The home was a grand, historically important home in a chic district of London. Heritage restrictions meant that the outside look of the house couldn't be changed, and many of the internal details had to be restored to meet with the approval of the local authority. The house was in a dilapidated state before its retrofit and needed extensive modernization and modification to improve its energy efficiency and install renewable solar technologies. Researchers from the University of London are now monitoring the house; the forecast reduction in energy use and carbon emissions is 82 per cent. University College London is carrying out ongoing monitoring of how the building performs in the field. Its analysis,[16] undertaken a year after the construction was completed, shows the building outperforms expectations. Losses of heat through ventilation have fallen by three-quarters, putting it among the top few per cent of UK homes. Heat loss through walls, windows and doors is better than that demanded by new-build regulations.

Over 300 people viewed the home in the first six weeks it was open, including visits from central government departments and many London-based community groups. People appreciated the opportunity to seeing modern technologies being applied to the type of old Victorian home many aspired to live in. The publicity surrounding the home hasn't all been good; the cost of the refurbishment was much higher than many visitors considered reasonable – around £60,000 for the energy efficiency and micro-renewables (the house is very large) – and there were some teething issues with some of the technologies. But the house provided an inspiration for others to think about whether their own period homes could be improved. Many of the visitors commented on the quality of the finish of the wooden sash windows and the way in which the cornices had been fitted on top of internal insulation. The reasons for visiting were varied: some were nosey neighbours, others had a spare hour or two to kill, but some had a sincere and sophisticated interest in whether it was possible to retrofit a heritage home while maintaining its style and – let's be blunt – its financial value. For most homeowners, their home isn't just the place they live, it's their main store of wealth.

Other community groups are also establishing exemplar homes. In Kitchener-Waterloo, Ontario, we saw the charity Residential Energy Efficiency Project (REEP) undertake low-energy retrofits to two homes that seek to shave 90 per cent off energy bills.

People trust the advice of competent builders and contractors. They also trust the advice and recommendations of organizations such as local government.[17] Doctors and other medical staff are highly trusted. This trust is utilized and surgery staff provide advice to elderly patients in winter about the threat to their health from cold and damp homes. The staff offer advice to patients on where to get more information about insulating their homes and reducing their bills.

Rewarding Good Behaviour

Another simple behavioural tool is rewarding good actions. When we're rewarded for doing something, we're likely to repeat it. One obvious reward is monetary gain from improved energy use, but this isn't always enough. The simple psychological reward from competing and winning is considered a key to the success of the Green Streets and Smackdown competitions. Other rewards included free or low-cost assessments, energy-savings devices such as CFLs and prizes such as free dinners. The most significant reward offered in the British Gas Green Streets competition was £100,000 for the winning team to invest in a community project. Obviously, this amount of cash provides a strong incentive.

One thing common to each of these examples is that they entail social engagement – people interacting with each other in groups. While it is possible to change the behaviour of someone in isolation, groups facilitate individual change. Why? First, groups are critically important to us. We are, in essence, social creatures. Groups provide us with the things we need – social support, companionship, a sense of belonging. Our notion of self is determined, in part, by our 'social identities' or the groups to which we perceive ourselves as belonging. Our self-esteem is very much affected by these identities.

Community Decision-Making

NIMBY *[Not In My Back Yard] n. Slang. One who objects to the establishment in one's neighbourhood of projects, such as incinerators, prisons, homeless shelters or wind turbines, that are believed to be dangerous, unsightly, or otherwise undesirable.*

Variations:

NIMTO *[Not In My Term Of Office]*

BANANA *[Build Absolutely Nothing Anywhere Near Anyone]*[18]

Many energy-saving technologies cannot be deployed by individuals. They have to be deployed by communities. In Chapter 5 we talked about some of the institutions that can help communities reach decisions. In this section we look at some of the processes of how communities come together to reach a collective decision. The economics of a DH system need a high proportion of people to connect to the network and stop using their existing heating fuels when the network is commissioned; external insulation of a housing or commercial block needs people benefiting from the technology to agree to it and fund the investment. Not everyone will like this decision. People disagree for a whole host of reasons: self-interest, aesthetics and differences in views on 'facts' :'wind power/double glazing' (*delete either or both according to your prejudice*) doesn't work. Virtually every new energy technology attracts criticism from people who don't like it, who have their own better ideas and who disagree with it as a matter of principle or as a matter of taste.

As pointed out earlier, wind energy is the second most popular form of energy in the UK after solar. People regard it as preferable to all forms of fossil fuel and nuclear fuel and yet this view is not uncontested.

The definitions above show there is a whole lexicon of terms for those who wish to prevent or stall new developments. As the UK's former Deputy Prime Minister John Prescott tactfully remarked at a wind energy conference:[19]

Time and time again we see ambitious and worthy wind farm applications defeated by a vocal minority of landowners and NIMBYS … did these people campaign against the mobile phone masts that allow them to call their locals up to organize the protests? Did they moan about the pylons that bring in electricity to their hamlets to power the computers that send out emails to lobby the councils against wind farm applications?

We, like many environmentalists, believe that planning processes in the UK and US give too much emphasis to the views of a vocal minority of people, against the (muted) support or at least acquiescence of the majority. At no financial cost to themselves, objectors can petition against minor external changes to the look of homes, making the process of improving the energy efficiency of a home more expensive than it need be, and increase the time and expense of building renewable energy schemes.

Why is this a problem? Don't objectors also try to prevent large centralized power stations and transmissions lines being built? Indeed they do, but the resources of protestors are small compared to the hundreds of millions that are spent on large infrastructure projects. They have become a cost of doing business, an annoyance but not a fatal impediment. The same is not true of small community-scale renewables where the cost of getting through planning can ruin the economics of the scheme. The reason why wind turbines are being developed offshore in the UK rather than onshore is partly because of the more reliable winds, but more because of the reduced cost of planning. In 2006, a major review of planning, the Barker Review, found that there was three times more capacity of wind turbines in planning than had been built, and that consideration of the applications took an average of ten months.[20]

It doesn't have to be this way. There are ways of discussing local energy planning that have had the poison taken out of them. Sustainable Hockerton, or SHOCK, is a village-level community initiative in the village of Hockerton, Nottinghamshire. It grew out of Hockerton Housing Project, a small community of five highly energy-efficient homes on the fringes of Sherwood Forest. The five houses are of great interest to low-energy connoisseurs in their own right. They consume just 20 per cent of the energy used by standard homes. The floors, walls and ceilings are cocooned in 30 centimetres of polystyrene insulation, which is then covered in earth to blend into the landscape. The south face of the terrace consists of triple-glazed windows, leading on to a sunspace that is in turn double glazed in order to capture the warmth of the sun. The homes are so well insulated and make such good use of solar energy that they do not need central heating.

In keeping with the ethos of the housing estate, the owners wanted to generate electricity renewably; wind was the most appropriate technology at the site. Rob and Liz, owners of one of the homes, explain how the rest of the village was initially hostile to the site's plan to put up a wind turbine: 'When we built the first wind turbine – just a small 5kW unit – it took us five years to clear planning. We decided we needed a better way of working with the rest of the village'.[21] In 2006 the housing project leafleted each of the 55 houses, pub, industrial estate and farms, to invite them to a meeting to look at ways of reducing the village's carbon footprint. Forty-six people from the village attended a series of meetings and workshops. Together they identified a number of ways to make the village more economically, socially and environmentally sustainable. They decided to build a second much larger turbine, enough to meet the entire village's energy needs. The second planning process took just eight months. According to Liz:

Only four of the fifty four homes in the parish opposed the plan, many more locals are investing their savings in the project. They are expected to make a decent return and the farmer whose land we are using gets a share of the profits. I think it's about More Haste Less Speed – time has to be invested in building relationships, with incremental change. People need to see exemplar homes or renewable energy systems in operation to reduce the fear of change.[22]

Bottom-up village hall activism is great for small decisions, where relationships can be fostered between all the affected parties. Decisions that can be taken in a community hall are decisions that impact on the people in the room within the next few years. But a different forum is needed for large energy infrastructure that affects entre towns or cities, where hundreds or even thousands are affected. Decisions about whether to commit a city to constructing a large power plant or a DH network are taken infrequently and the consequences last for many decades. How should affected parties be polled so we get a representative view? The current way of doing things – inquisitorial planning inquiries about specific projects – is a good way of hearing what people oppose; it is not so good at asking people what they support and how they would trade-off different public objectives.

Blair Hamilton from Efficiency Vermont speaks of how people in Shelburne, Vermont, found themselves in a dilemma. The town's people objected to the construction of new transmission lines to bring more power into the town. Some townspeople were opposed to wind, others to fossil fuels. But everyone wanted the use of electricity. How could they navigate through this maze of disagreement without destroying the community? Shelburne formed an Energy Committee to try to implement the preferred vision of more energy efficiency, energy from farm waste, geothermal and solar hot water.

A similar challenge, but at a much larger scale, faced the entire state. Vermont has decided not to relicense its 540MW nuclear power plant. The plant represents three-quarters of the state's generation capacity, so the decision leaves a huge gap in its capacity to meet its power needs within state (something the state treasures). The state was interested in finding an effective and democratic means of interrogating local preferences (see Box 8.6), but most members of the community aren't interested in participating in the sorts of forums where such decisions are taken – that's why they vote for politicians. In autumn 2007 the state's government asked the University of Texas to undertake a two-day deliberative poll.[23] The 150 Vermonters were picked at random and spent a weekend listening to evidence from all sides of the argument.

Results from such a considered process seem to us much more valuable than the current system of planning that gives too much power to those with trenchant views and too little emphasis to those who have a view, perhaps a silent majority, but do not find the process of inquiry to their liking. Worse than this are the current processes for engagement with community that give people the opportunity to voice opposition but too little chance to signal options they would prefer or specify trade-offs between factors.

It's one thing asking a community about their preferences. It's another thing making the decision stick. Do local decision-makers have the necessary powers

Box 8.6 Vermont's energy future

Over a year the state government undertook a project to find out exactly what Vermonters think about possible future sources of electric power. With contracts for Hydro Quebec and Vermont Yankee nuclear due to expire in a few years, where should Vermont get its power after that?

One weekend, 150 Vermonters were selected at random. They met for two days at the University of Vermont to discuss and learn about electric power options. The panels called witnesses from the energy companies and all the major different sources of power. The participants were randomly split into 13 small groups. After the deliberations the participants wanted to see a quarter of the energy to come from hydro, 18 per cent from wind and around 15 per cent from solar, wood and nuclear, in that order. They wanted almost none of it, however, to come from oil or, especially, coal. Just over half wanted the nuclear power station to continue in operation. Interestingly 90 per cent were in favour of wind turbines being built even in places visible from where they lived. The process of deliberation did not greatly change people's existing views; however, it reduced the amount they believed could be saved from energy efficiency from a third to 22 per cent. Deliberation significantly increased the sought-after share of hydro (imported from neighbouring Quebec) and decreased that from solar because of the increase in bills. The process provided decision-makers with a genuine insight into the informed views of people where they have to make choices based on their values, the relative costs of their preferences and the discussions with others who might have different views to their own. Richard Watts helped put together the event. He says, 'There's this constraint in Vermont between taking what people clearly care about and what actually emerges as the public policy outcome'.[24]

Nuclear has been controversial for decades, drawing periodic protests. The poll found a nearly 50:50 split in support versus non-support. Dottie Schnure of Green Mountain Power Corporation says that while nuclear remains divisive, the idea of small renewables is popular: 'Our polls that we have done show that people are interested in more renewable and small generation ... People say they're interested in wind. It's been difficult to locate wind plants'.[25]

The poll showed much support for wind – even if a project were visible from a person's home. The people who were polled indicated that they would be willing to spend somewhat more for renewables, even though wind power in particular is not as reliable when the wind stops blowing.

But Richard Watts says the preference for wind is a mandate that should not be ignored: 'It's clear that the state needs to do more to support large wind facilities in Vermont ... That's one of the clear messages I think that should be taken from that weekend of asking Vermonters what they think'.[26]

The question is: will future energy policy actually follow the results of the poll? That is, will electric utilities move away from large out-of-state sources and rely more on small, in-state renewable sources? It's a question that still awaits an answer.

to make their community comply with the agreed path? Sometimes people that don't support the decision have to conform to the will of others. In the UK, local government councillors interested in improving energy use within their communities might look wistfully at cities in the US that can create brand new regulations to mandate compulsory energy efficiency improvements to existing homes, like the RECO ordinance in Berkeley, while they lack the power to even forbid a restaurant or café from using gas burners.

The 1970s were a traumatic decade for energy planners with two oil shocks in 1973 and 1979. Denmark found itself on the wrong side of the vastly more expensive global energy trade. In the mid-1970s, realizing how dependent it was on foreign oil imports, it began to plan on how it might improve the economy's energy efficiency. But it was the 1979 price rise that gave the country the political cover to introduce the Heat Law and enshrine a much more interventionist role in energy supply. The nation chose to localize the development of energy plans. The Heat Law asks local government to plan heat provision within its neighbourhoods; local government decides which streets should use DH or individual central heating from gas or some mixture. Interestingly, the UK at the same time made the decision to drive through a central government-led roll-out of gas networks.

In DH areas, the authorities in Denmark had to identify streets for collective heat networks and potential sources of heat. Municipalities were then responsible for providing the heat load either from existing heat networks or developing new ones.[27] Once built, citizens were required to connect up to the facility within nine years and could be prosecuted for not doing so. The rules were quite draconian – electric heating was banned in new buildings and the use of gas heating was banned in heat network areas. These bans nakedly helped the economics of the network by making it the monopoly supplier. It reduced individuals' freedom to choose but also reduced the cost of financing the heat network because lenders could see that the income stream was guaranteed and their loans would carry a low risk of non-repayment. No DH company has defaulted on interest payments. How did people respond to having to compulsorily connect to the heat networks? There were losers like people who had recently bought new oil boilers, which they had to scrap. Birger Lauersen who works for the Danish DH association explained that by and large people were happy – the country understood the difficult position it was in regarding energy and the nation showed leadership in explaining the need to switch to DH:

> The cost of connecting to the district heat was typically around £5000; this covered the pipework between the network and the house, the purchase of the heat exchanger and the meter – which is quite a lot of money but people accept this, because it became cheaper to heat homes. People living in a street used to get together and ask if the district-heating network could be extended to their street. Because individuals had up to nine years to connect it meant it was no a surprise. We also provided grants to poorer areas in the late 1980s to allow them

to connect. The technology was also already well known in Denmark since around a third of homes used district heating even in the 1960s. It was regarded as a very reliable technology, with few problems for users.[28]

He explained that the system in his home provides warm water at 65°C, which is used in his central heating system. The meter is remotely read every month – the garbage truck that drives past his home logs the data from a small radio broadcaster in the heat meter. His supply company also provides power. Did he not mind the fact that the company was a monopoly and he wasn't allowed to switch? 'There has only ever been one interruption in the heat service. The bills are on time and not too expensive. That's all I care about.'

To protect consumers from exploitation by the monopoly heat provider and reduce hostility to the mandatory changes, the government simultaneously put restrictions on the heat supplier. No profit could be made from heat sales, and the heat supplier had to maintain open books to demonstrate this to customers. In practice only municipalities and co-operative companies have chosen to enter into the heat supply business.

In countries with liberalized energy markets, the creation of a regulated monopoly seems a backward step. But now heat customers enjoy prices that are 30 per cent lower than gas customers and communities clamour to have the heat network extended. Denmark has the second lowest energy use per unit of GDP (after Switzerland) in Europe and North America. Progressive heat co-operatives employ digital heat metering and let clients manage their account on the internet. The most modern CHP power stations claim a 40 per cent improvement over the separate production of heat and power, once account is taken of the low line losses and successful use of most of the waste heat from thermal power production.

By almost any criterion this rejection of market liberalization and replacement by powerful local planning and compulsion has been a success. But few countries have dared emulate this rejection of conventional wisdom and allowed local government to choose and implement the development of its own preferred infrastructure.

Changing Installers and Products

It's not just customers that need to change if we are to get widespread deployment of energy efficiency and local generation. One of the reasons that consumers are reluctant to spend their savings on energy efficiency is because they are worried the measures won't work properly, won't be installed properly or will end up causing more problems than they solve. They have good reason. Most people who have tried to install energy efficiency measures in the home will have had mixed experiences of good and bad quality work, reliable and unreliable builders and inexperience or lack of skills in contractors. Prashant has documented some of his own experiences in his earlier book *The Economical Environmentalist*.[29]

Improving consumer confidence and hence uptake means communities have to develop effective ways of increasing the skills in the industry. Many of these skills are nothing to do with energy efficiency – simple things such as improved punctuality, providing realistic quotes, restoring the site after the work is complete, effective redress when things go wrong and ideally some kind of industry-wide assurance scheme so that guarantees will be honoured even if a particular business ceases to trade. Consumers also need to be able to easily discriminate between rogue businesses and those that meet their promises.

Larry Whitty was the minister responsible for energy efficiency in the UK and he introduced regulations mandating the use of gas condensing boilers and requirements for energy utilities to install energy efficiency. In interview, he told us: 'It was not easy to take industry with us. They made rhetorical commitments – in reality the building industry so that not interested in regulations that changed their existing practices. They were interested in more work, but not more complicated work.' We asked if the issue was about reskilling and Whitty replied:

> That was a large part of it. The workforce were used to a certain way of doing things. In Defra [Department of Environment, Food and Rural Affairs] we saw gas condensing boilers as the most cost-effective opportunity. But introducing regulations to make these mandatory met with deep resistance from the trade associations. Defra facilitated a skilling programme – through the colleges – we needed to get 40,000 fitters through the course before the regulations came in. It was actually only a few days of training but it required working across departmental boundaries and the costs were mainly borne by industry. We began arranging this two years before the regulations went live.

In the US, Energy Star run a Quality Installation programme to train people to ensure heating and ventilation equipment is properly fitted. Quality installation means sizing the equipment correctly (often installers fit heating and cooling units that are too large for the building, which as a result run inefficiently), sealing the ducting properly, setting the air-flow rate and filling with the correct amount of refrigerant. All of these are invisible to the customer, all of these can and are done wrongly. According to Energy Star, half the heating and cooling equipment never performs to its advertised capacity. Efficiency Vermont encourages installers to go through the training scheme. Blair Hamilton explained to us:

> Energy Star QI [Quality Installation] has been around a long time so contractors don't mind going through it – it is not seen as a 'flavour of the month'. We subsidize the contractors so they don't have to pay the full cost of the five-day training. So if the cost of course is $1000 we give back $600 back once the contractor completes the course and does a minimum of two jobs in the field. We also subsidize the purchase of tools – as these can be expensive.

In Sacramento, SMUD maintains a list of pre-approved contractors. As well as giving consumers greater confidence in adopting the technology, these contractors act as a marketing arm for their work. Ed Humzowee explained to us that to get onto the list, SMUD undertakes some prequalification checks on the contractor: the contractor has to provide a two-year warranty on his work and have all the appropriate licences. To carry out a Home Performance programme, the contractor requires Building Performance Institute training. So large a proportion of the city's contractors are now on this list that the job of keeping the database up to date (checks are refreshed annually) has become quite onerous.

Building greater confidence also requires enforcement to pick up and rectify poor practice. Building inspection is the Cinderella service within local government and is chronically under-resourced. This can and should change. Building inspectors can drive change within their communities by ensuring poor quality work is spotted and rectified. Larry Whitty speaks of an inspector that visited his home to examine an extension built soon after an energy efficiency rule change; he was pleasantly surprised to find the inspector informed him of the new standards and spotted deficiencies in the contractor's workmanship that had to be rectified.

Inspectors should go further and adopt severe sanctions against builders, for instance 'three strikes and you're out' type rules if contractors persistently construct substandard work. It is important that inspectors have no conflict of interests. In some parts of the UK, local authorities allow large commercial buildings to self-certify their buildings; in other areas the client is allowed to select the inspector, usually taking the advise of the builder, creating a hopeless conflict of interests.

Some of the best rates of return are in improving energy use in commercial building. In Toronto, Enerlife maintains a database on the energy consumption of individual commercial buildings as part of the mayor's Megawatt Challenge. Large commercial buildings typically have dedicated facility management staff. Enerlife's Noel Cheeseman explained to us the company's philosophy:

We are obsessive about the performance of the existing building stock. We show and probe the data by sector. Our core business is to provide benchmarking data to property owners and managers to achieve and demonstrate high standards of energy and environmental performance in their individual buildings. Our main metric is total energy per unit floor area.

He believes most facility managers grossly mismanage large commercial buildings in Toronto. He wants to help make building owners better customers, ensuring that they avoid purchasing unnecessary or oversized equipment.

Toronto's tower renewal challenge programme is aimed at the city's 1000 high-rise buildings. Issues found in social houses include poor heating controls and a lack of individual meters for each household's gas supply. Noel Cheeseman worked with the social housing corporation and showed them how to identify wasted energy: 'Take a look at a building during a really cold winter. If you see windows open at the top there's a stack effect going on. If the door whistles when you open, same thing. You can reduce emissions by 10–20 per cent by addressing this.'

Seeing the variation in performance across similar buildings convinces him of the scale of the challenge, but also of the opportunities: 'The average energy use for similar buildings could be double that of the best performers. Often small tweaks in the system improve performance: for instance a minor component in the ventilation system might not be working, or equipment might have been wrongly installed.' Laughing he told us how he once found a fan in a cooling system that had been installed backwards. Building managers can also automate or zone circuits to ensure that banks of lights and printers are switched off en masse when not in use. Beyond this he shows building managers how checks are made to ensure existing equipment, such as thermostats and motion-detecting sensors, are correctly operated.

Some energy-saving opportunities should not be localized. National governments, the EU Commission and the federal government have to act. Appliances, electronic goods, air conditioning equipment, energy meters and, soon, electrical vehicles are designed and built for regional markets, and they should be regulated at this level too. Already these goods are regulated so that they meet minimum energy standards; energy labels identify goods that perform to high environmental standards.

The use of minimum standards is an example of government carrying out *choice editing*. We think it should do this more, where it is self-evidently in everyone's interests (apart from recalcitrant manufacturers) to set appropriate minimum standards. Minimum standards are particularly well suited to hard-to-observe characteristics such as maximum power use of gadgets while in stand-by mode – the IEA began to talk about the 1 watt plan in 1997. The cost to manufacturers of complying with this very sensible rule is almost nothing, and every watt of continuous power consumption uses 9kWh a year and costs around $1. How many gadgets are sitting around the typical home wasting energy for no good reason? Instead of behavioural change and education programmes, it makes more sense simply to ban such energy waste.

Energy labelling is a great idea but there's room for improvement. In the EU it has been compulsory to display A to G energy labels on appliances close to the point of sale so that the customer has unbiased and comparable information on the energy efficiency of goods. The A to G message is used throughout the Union and for every product. The labels use red, green and amber colours to amplify the message. Most consumers in Europe claim to take account of this information (for instance 58 per cent of consumer claim to pay 'a lot of attention' to a fridge's energy efficiency) when making a decision about which model to buy. While people use energy labels, most don't spend much time thinking about whether they really are buying the best products for them. In a way, the labelling scheme is now a victim of its own success. Manufacturers have largely removed fridges that are less than A-rated from the market through better design or changes in stocking practices. But instead of revalorizing the labels so consumers are still able to discriminate between the best and the worst performing, the industry and the Commission have spent years squabbling about how to go forward. Also the labels refer to relative rather than absolute energy performance. This means that a small C-rated appliance might actually use less energy than a large A-rated appliance; something a consumer might well not appreciate.

In the US, the Energy Star label is intended to identify the best 25 per cent of products within a good of that size. Here the standard is kept up to date, in theory. According to Rich Brown at the Lawrence Berkeley Lab it takes around 12 to 18 months for a new test to be agreed, so in fast-moving sectors the products have already moved on. For the most part, the rating is relative rather than absolute, so a starred product is among *the best for its size*. Interestingly, Energy Star has decided to base its recommendations for rating TVs irrespective of size – which we think makes much more sense.

Recommendations

For a long time we have treated behaviour towards energy-use change as though it were a failure of education or awareness. If changing our behaviour was all about imparting public information then we should all give up now. The other side has too much money. The environmental communications consultancy Brook Lyndhurst undertook a study of the advertising campaigns that took place between 2006 and 2009 in the UK.[30] Of the 600,000 advertisements on file, just 4000 made any sort of green claim or had a green message, and of these just a fifth were public information trying to change our environmental behaviour. The vast majority of paid-for advertising is about increasing our use of goods and services, or shifting them from one company to another.

The behavioural change we are talking about is much more difficult to tackle as it is more ingrained. But as a society we have been successful in this sort of behaviour change before, particularly in the field of public health – the rate of smoking in North America and Western Europe has been falling dramatically through a mixture of taxation (high tobacco taxes), regulation (preventing advertisements glamorizing smoking and compulsory labelling) and health education. Consumer sentiment towards smoking has changed so much since the 1950s that employers banned smoking from most offices even before it was outlawed.

There is a second sort of behavioural change needed if we are to deploy energy efficiency and distributed generation. This is rethinking the way we assess public attitudes about changes that we need to make. Local leadership is not about caving in to the vocal. In the UK the development of onshore wind is being halted by objections by a minority of people, using the planning process to slow down and greatly increase the cost of deployment. Minor changes to the outside of 'heritage' buildings are also made bureaucratic and unnecessarily expensive, deterring many from ever considering modifying their homes.

This can be seen as a triumph for local preferences. But it isn't really. Poll after poll shows that the majority of people living in the vicinity of new wind plants support them. Those campaigning against change are a minority. In the UK, 70–80 per cent of the public support wind farms, and only around 10–15 per cent oppose them. The figures barely flutter when the question refers to a wind farm being erected near their home. Support is highest in communities that actually have a wind farm nearby. Familiarity breeds content, not contempt.

The same conservatism is holding back cost-effective energy efficiency upgrades. Oxford University's Environment Change Unit estimates that 5 per cent of UK homes are in conservation areas and around 370,000 are listed by the national heritage authority, which means that any changes to the inside or outside of the building requires the consent of the government.[31] In the US, programmes to retrofit old homes using stimulus money are being held up while the conservation lobby considers each application.

Deliberative polling gives a voice to the silent majority of residents. It allows local decision-makers to listen to the way they think about energy wastage, local employment, energy security and acceptable trade-offs. This is something local communities have to work through using local politics. But once a settled position is reached it has to be implemented through planning and rule-making.

Notes

1 Wilson, E. O. (1998) *Consilience: The Unity of Knowledge*, Abacus.
2 Leiserowitz, A., Maibach, E., Roser-Renouf, C. and Smith, N. (2010) *Americans' Actions to Conserve Energy, Reduce Waste, and Limit Global Warming: June 2010*, Yale University and George Mason University, New Haven, CT, http://environment.yale.edu/files/BehaviorJune2010.pdf.
3 European Commission (2007) 'Special EUROBAROMETER 262 Energy technologies: Knowledge, perception, measures', European Commission.
4 European Commission (2006) 'Special Eurobarometer: Attitudes to energy', European Commission.
5 Leiserowitz, A., Maibach, E., Roser-Renouf, C. and Smith, N. (2010) *Americans' Actions to Conserve Energy, Reduce Waste, and Limit Global Warming: June 2010*, Yale University and George Mason University, New Haven, CT, http://environment.yale.edu/climate/files/BehaviorJune2010.pdf.
6 Department for Environment, Food and Rural Affairs (2008) 'A Framework For pro-environmental behaviours', www.defra.gov.uk.
7 Thaler, R. and Sunstein, C. (2008) *Nudge: Improving Decisions About Health, Wealth, and Happiness*, Yale University Press.
8 Cabinet Office & Institute for Government (2010) 'MINDSPACE Influencing behaviour through public policy', www.instituteforgovernment.org.uk/images/files/MINDSPACE-full.pdf.
9 Friedrich, K., Amann, J., Vaidyanathan, S. and Elliott, N. (2010) 'Visible and concrete savings: Case studies of effective behavioral approaches to improving customer energy efficiency', American Council for an Energy-Efficient Economy, Washington, DC.

10   See www.energysmackdown.com.
11   IPPR (2009) 'Green Streets', www.ippr.org.uk/uploadedFiles/research/projects/Climate_Change/green_streets_final.pdf.
12   Allcott, H. (2009) *Social Norms and Energy Conservation*, Centre for Energy and Environmental Research, MIT Energy Initiative, Cambridge, MA.
13   Hopkins, R. (2008) *The Transition Handbook: From Oil Dependency to Local Resilience*, Green Books Ltd, Totnes.
14   National Audit Office (2008) 'Programmes to reduce household energy consumption', HC 1164 Session 2007–2008, House of Commons.
15   Prashant's interview with John Doggart, May 2008.
16   Ridley, I. (no date) *The Performance of the Low Energy Victorian House*, University College London.
17   ICARO Consulting (2009) 'Understanding consumer attitudes to sustainable community infrastructure', UK Green Building Council, www.ukgbc.org/site/resources/show-resource-details?id=601.
18   Our definitions.
19   British Wind Energy Association Conference, 20 October 2009.
20   Barker, K. (2006) *Barker Review of Land Use Planning*, www.communities.gov.uk/documents/planningandbuilding/pdf/154265.pdf.
21   Interview with Liz Laine, September 2010.
22   Interview with Liz Laine, September 2010.
23   Luskin, R., Crow, D., Fishkin, J., Guild, W. and Thomas, D. (2007) 'Report on the Deliberative Poll® on Vermont's Energy Future', University of Texas at Austin, www.cdd.stanford.edu/polls/energy/2008/vermont-results.pdf.
24   RAAB Associates (Nov 2007) 'Vermont's Energy Future: Regional Workshops', www.raabassociates.org/Articles/VTENG%20Final%20Report%2011-27-07.pdf. Watts comments on www.wcax.com/global/story.asp?s=8017484.
25   WCAX.COM (2008) www.wcax.com/Global/story.asp?S=8017484&clienttype=printable.
26   RAAB Associates (Nov 2007) 'Vermont's Energy Future: Regional Workshops', www.raabassociates.org/Articles/VTENG%20Final%20Report%2011-27-07.pdf. Watts comments on www.wcax.com/global/story.asp?s=8017484.
27   IEA (2008) 'Denmark – answer to a burning platform: CHP/D', www.iea.org/g8/chp/docs/denmark.pdf.
28   Prashant's interview with Birger Lauersen, 5 November 2010.
29   Vaze, P. (2009) *The Economical Environmentalist*, Earthscan.
30   Lyndhurst, B. (2009) 'Assessment of green claims in marketing', Report for Defra, www.brooklyndhurst.co.uk/media/1c99f4e9/EV0430%20-%20Assessment%20of%20Green%20Claims%20in%20Marketing%20-%20Summary%20for%20Policy%20-%20Final.pdf.
31   Boardman, B., Darby, S., Killip, G., Hinnells, M., Jardine, C.N., Palmer, J. and Sinden, G. (2005) '40% House', www.eci.ox.ac.uk/research/energy/downloads/40house/40house.pdf.

Chapter 9

Bringing This Together – A New Ecology of Energy Markets

What Needs to Change

According to Darwin's Origin of Species, *it is not the most intellectual of the species that survives; it is not the strongest that survives; but the species that survives is the one that is able best to adapt and adjust to the changing environment in which it finds itself.*

Leon C. Megginson[1]

Large-scale problems do not require large-scale solutions; they require small-scale solutions within a large-scale framework.

David Fleming[2]

We started this book dreaming about a future that we hope might come to pass. But there is a large distance between our hopes for energy policy and what is actually happening. In Charles Darwin's world, organisms compete against one another in an endless kill-or-be-killed struggle. The idea of contest or of competition seeps throughout thinking about energy. Energy suppliers steal customers off one another by offering lower tariffs. Generators undercut one another moment by moment to tempt the system operator into dispatching their power. Low-cost fossil-fuel-exporting countries set the floor price for oil and gas markets globally. Strong suppliers drive small suppliers to extinction.

Large international energy companies, which generate, transport and supply energy, now dominate the energy market. In the UK the shorthand phrase we use to describe the strongest energy companies, the Big Six, echoes the pantheon of trophy animals caught by East African hunters. But the energy market is not a forum for Darwinian struggle, and the large companies do not dominate because of their intelligence or vitality. Each of the contests listed above is rigged. Consumers rightly distrust suppliers because they find that their energy price is raised shortly after the marketing campaign that persuaded them to switch is concluded. The suppliers/generators have shaped the energy market so it suits them commercially. Most politicians listen more to big energy companies than to consumers, scientists or society at large.

We probably wouldn't worry about the flaws in these Darwinian arenas of contest if society's demands from the industry were static. But society's demands are not static. The pressures of climate change, energy security and maintaining affordability mean things have to change. So public policy takes these arenas and overlays them with carbon markets, complex price mechanisms, information campaigns and regulation to incentivize, educate and compel businesses and individuals into developing more sustainable energy patterns.

Despite these policy interventions, energy policy is failing, by almost any criterion. Wholesale energy prices are volatile. Dependency on imports from unstable countries is growing (though discoveries of unconventional gas reserves might add another decade or two to our reserves). Emissions of CO_2 in North America and northwest Europe have barely fallen over the last decade, despite Kyoto-style carbon markets. Energy is becoming less affordable for low-income households. While many countries have succeeded in expanding the share of renewable electricity, this has only happened because of generous financial support, not all of which will be affordable to taxpayers or energy customers. CCS and new nuclear have failed to make significant in-roads in any Western economy yet. Development of new DH is at a standstill, despite some people's belief that it is the only viable option for heating old, high-density city centres. Impatient capital is uninterested in investment in such long-term assets unless the returns are much higher. And no government has a credible programme about how homeowners and commercial premises might be induced to pay for the retrofit of their buildings.

The large incumbent firms with their huge sunk costs – not just in terms of physical capital but intellectual capital – are ill-suited to carry out the necessary innovation. This should not be read as a criticism of the people working in these firms. Both of us are close enough to these businesses to know there are clever and well-meaning people working in there, including at the most senior levels. But there is too much to be gained commercially in keeping things just as they are. And footloose global capital does not respect innovation just because it is well intentioned. Capital requires innovation to make a return, preferably in the next quarter.

Back in 2004 The Economist carried an article about the 'Energy Internet', seeing parallels in the distributed, bottom-up, vaguely anarchic manner in which the internet developed and the way the smart grid may yet develop.[3] Web 2.0 means that anyone with a personal computer and an internet connection can now contribute content to the web. Across the world, untold servers are now linked together. No one is in charge but the glue of protocols and conventions make this free-for-all somehow work. And there is discipline despite the fact no one is in charge: mirror sites carry faithful copies of the most important sites to ensure data are not lost, googlebots trawl around the web caching indexes to ensure speedy data retrieval, distant server banks store the internet protocol addresses of all the computers that hold content so that browsers may find material.

This book argues for much more local production of electricity and heat. We believe energy systems should operate much more like the internet, in the sense that energy flows should become more distributed and less centralized. There are efficiency benefits from local energy generation: easier utilization of waste heat as production is brought closer to major centres of heat demand, lower transmission and distribution losses, reduced need to strengthen the transmission and distribution lines. There are lower risks: avoiding the threat of a single large power station or power line falling over. A more distributed system operation means there is less redundancy in the overall design. Our present system has substantial overcapacity in transmission and distribution networks, to ensure that voltages and power quality are always safeguarded. Intelligent local grids can moderate and smooth demand, for instance by storing electricity in electric vehicles or using distribution grids to mobilize small-scale CHP and smart appliances depending on the instantaneous generation of renewable energy.

Such a shift would be the equivalent of the shift in biology from a Darwinian to a Margulisian viewpoint. In the 1970s Lynn Margulis argued that early bacteria co-operated and formed a deep association with large-celled organisms.[4] This led to benefits for both parties, and allowed the evolution of multi-cellular organisms such as humans. Every cell in our body is now equipped with its own bacterial fuel cell, mitochondrion, which provide the cell with ATP (adenosine triphosphate), the biological equivalent of electricity. This understanding of the ancient act of decentralized co-operation at the sub-cellular level is causing a rethink in biology, with a greater appreciation of co-operation between species, rather than raw competition.

There are intriguing parallels for energy. Instead of relying on competition between oligopolistic firms to provide sustainable energy systems under government mandates, we argue that local communities can play a much more central role in creating, storing and intelligently using energy, but still relying on networks to link the different islands of generation.

Like the creation of the internet, this will require a new corporate etiquette to work. Local, non-profit-oriented organizations are more likely to gain the trust of communities and more likely to teach, inspire and lead them to lower energy-using behaviours. We hope we have shown that distributing the responsibility confers greater responsibility on consumers and demands greater innovation. Communities in Upper Austria, Denmark and Vermont see energy as a means of delivering broader community goals: rural employment, energy independence and affordable warmth.

We think that companies driven by shareholder returns, the never-ending zero-sum grab for market share, the dark pallor of vertical integration and the nation-straddling ambitions are wrong for the task ahead. We argue for co-operation within communities rather than competition between corporations to deliver energy efficiency and community energy. Energy production should be under much more local control than we see now.

And make no mistake, the two systems are at odds. Energy efficiency and distributed generation compete with the centralized production of gas and power. There is no point in pretending there are significant synergies. By and large, the more energy that is saved or the more energy that is produced locally, the less demand there is for centralized electricity sales. Substantial reforms are needed to permit community generation, local demand management and heat production to compete on an equal footing with centralized production. Centralized electricity companies will not easily allow the market rules to be amended to permit small new-entrant generators. Change would undermine their lucrative ability to benefit from profitable activities such as energy trading, balancing and regulating the market. But it is possible – indeed essential – for governments to make the changes.

It's Complicated

Everything big is made of small parts. The mighty desert is just grains of sand. The grain stands and says I'll make a difference. Soon there'll be others lending a hand.

'Small grains' by Jess Gold, 2010[5]

A friend of Prashant's told him of how his daughter had recently changed her Facebook relationship status from 'Single' to 'It's complicated'. The friend's response, 'Jesus wept', was understandable – his daughter is after all only ten years old.

This sentiment has its echoes in energy policy too. About the only thing that most energy experts agree on is that a single instrument is not enough, and that it's complicated. 'Utility-only cap and trade no silver bullet',[6] 'Exclusive Mercedes interview: Electric cars are not the silver bullet'.[7] Energy policy is complex and will remain so for the foreseeable future. If we are to simultaneously encourage individuals to fit solar thermal to their roofs, and companies to spend billions researching and fitting CCS, it is inconceivable the same policy will be sufficient to incentivize both.

The aphorism 'No silver bullet' recognizes a broader truth. We must make progress and succeed along many different fronts. Numerous policies need to be navigated through legislatures, many different technologies have to be deployed, many different actors throughout society have to be influenced and mobilized.

This last chapter of the book details our view of the policies and actions we think are needed to repower communities. We have not tried to map out which energy technologies will prevail. We don't know. More importantly, we do not think that these decisions can or should be made by policy wonks sitting in the centre.

The decisions about what works have to be decentralized. Some are a matter of aesthetics and local cultural preferences, others have to respond to the hard physical and climatic realities of the host area and others again are contingent on unknowable events, which will unfold over time. What we do is simply list policies to develop the circumstances where communities will reduce their energy use and develop clean energy.

Our Recommendations

Figure 9.1 summarizes our ideas, which are discussed in detail on the following pages.

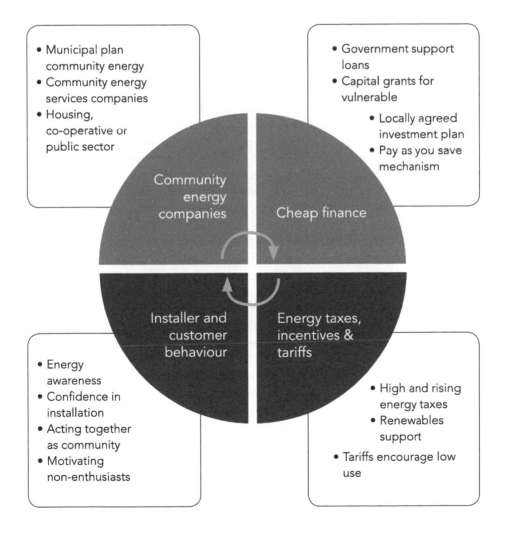

Figure 9.1 Bringing it all together

Municipal or local government should be responsible for planning community energy and area-based energy efficiency.

Local governments, acting either individually or jointly with neighbouring local authorities, should plan how an area will reduce and decentralize its energy use. These plans should incorporate local ambitions for housing, industrial growth and economic development. Plans might specify the locations of gas, electricity and heat networks, area-based energy efficiency projects and distributed heat and power generation. This will require a process of consultation with citizens and community groups but must then reach a conclusion. Deliberative techniques should be used to gauge preferences and more importantly 'settle' the local trade-offs between issues such as heritage, jobs and affordable energy. If the country has set national carbon budgets, which we think all countries should, local governments ought to set their own carbon budgets to be broadly in line with national targets.

Why local government? Why not leave it to the market? Planning infrastructure is a community decision, and an intensely political decision. People will have opposing views and preferences. While most people agree on the need for secure reliable energy sources, and more energy efficiency, there is no consensus about what technology should be allowed or where it should be located. Even among environmentalists there are strongly held views opposing onshore wind as being a blot on the landscape, energy from waste due to its emissions, use of biomass for destroying rainforests, hydro schemes for disrupting the meanderings of aquatic mammals and external insulation for 'disfiguring' buildings. Some of these views might be ill-informed, sometimes comical, and often the anger of opposing voices is out of all proportion to their numbers. But a local democratic process is needed to resolve these sorts of matters and make a binding decision. A referendum is needed on the proposed plan because opponents of any change speak out much more often and more loudly than do supporters.

As well as being an arbitrator between interests, local government is also a substantial owner of schools, offices, community facilities such as swimming pools, and of social housing. Local leadership also means quite literally 'putting your own house in order'. By being an intelligent purchaser of low-energy products and civic buildings, local government can provide a platform for teaching local trades and developing the local skills base. It can also act as an anchor customer for heat from heat networks, greatly improving their economic viability.

Local government is also an important regulator. Its building and planning inspectors can enhance a community's confidence in low-energy solutions by enforcing regulations wisely. Inspection technologies such as thermal imaging cameras and pressure tests can be used to identify deficiencies, improve building practices and to demonstrate good practice and inspire confidence in customers. If local government doesn't do this, no one else will.

New community ESCOs should deliver energy efficiency;
new distributed generation infrastructure companies should
build new assets

We believe the vertically integrated international energy firms should no longer supply energy or energy services to customers. The experiences of the UK and other European countries have not been a success: the conflict of interest between the company's upstream ambitions and the needs of a sustainable energy system are too great, and our efforts to tame or obligate them are too easily outwitted. We need to migrate over to area-based monopoly energy service providers where not-for-profit, municipal or customer-owned organizations are the customer-facing entity, perhaps partnered with private companies separate from upstream energy production. We call these community ESCOs.

New assets: We also believe that new organizations should build the assets outlined in the local government's infrastructure plans. Local government should offer concessions or franchises to build and operate new community energy assets. Community-scale generation (biomass-, biogas-, or waste-fired CHP, local wind, solar or geothermal), new heat networks (smart grid, heat infrastructure) or storage (electrical vehicle recharge parks and large thermal stores) could be offered to tender. Waste heat from large thermal power stations could also bid to supply heat and any waste heat vented into the environment should be taxed.

This is competition for the market rather than competition within the market. The community ESCO would be required to purchase heat from the local producers. The heat market is likely to be a local monopoly and would have to be regulated.

Community ESCOs: In sparsely populated areas the community ESCO might grow from free-standing energy efficiency utilities such as Efficiency Vermont or supply-only energy businesses such as Danish heat co-operatives. Users or customers would be involved in the management of the organization. In cities – with more complex needs – local authorities might set up or enter into partnerships with private companies with the appropriate skills for billing and customer account management.

But this will not happen spontaneously, nor overnight. Public policy-makers need to intervene to allow the development of such locally accountable organizations. The energy supply divisions of the energy companies have many of these skills. We see merit in structurally separating these from the generation business. There is also merit in combining community ESCOs and ownership of local networks. This would allow the planning of smart grids that are easier to coordinate. For instance, decisions about how much to rely on recharge points for electric vehicles and customers' appliances to reduce demand for power would be made by the same organization.

Whatever the legal form of these organizations, they need to be constituted so that their mission is to reduce the use of fossil fuels and meet energy service needs. We would envisage that they pass on the wholesale cost of electricity and heat and taxes to their customers. The money the ESCO retains for itself should not be related to *the amount of energy it sells; it should be linked to the amount it saves.*

This could unlock tremendous innovation. Small-networked organizations have the potential to learn off one another. They already do so. Programmes such as Shade Trees or Smart Roofs run by the Californian municipally owned electricity company SMUD are innovative and popular ways of reducing air conditioning loads. The PACE mechanism for property tax bill financing was swiftly diffused through localities in the US through professional and informal networks. Many atypical sorts of firm are involved in community energy and energy efficiency already. These include building service companies, municipally owned companies, special purpose vehicles jointly owned by government and the private sector, customer-owned organizations and social housing organizations. New organizations might even develop from NGOs such as Transition Towns, rooted in the community. We need to give these organizations space to prosper.

Government must ensure low-cost capital is available to community energy developments

The energy industry will spend many hundreds of billions over the coming decade on a new generation of plant, transmission and distribution lines, and metering. Utilities can borrow money cheaply from the capital markets because energy regulators ensure that the repayment costs of approved capital projects can be recovered from energy bills. As a result, utility bonds in Europe are well regarded by the lenders and rating agencies.

Community energy projects cannot access capital on such favourable terms. They are less well understood by the financial community and their projects are often seen as 'unbankable'. Retail competition makes it more difficult for energy suppliers to invest in energy efficiency. Through much of Europe and some of North America consumers freely switch energy supplier, owners and tenants sell and move home regularly. How can an investor be sure of seeing a return on their investment with so cavalier a counterparty? Without a long-term contractual relationship these projects will be seen as risky.

Payment for long-term investment in community energy infrastructure has to be recovered from beneficiaries. One approach is to permit the repayment of investment, once sanctioned by a democratic body such as a local authority, to be permanently linked to an address or a meter so the investor sees a return, even if the supplier, building owner or tenant changes. There are many approaches to doing this. The best local solution depends on existing laws and customs. Vaze and Mayo (2009) suggest how in the UK the 'regulated asset base' could be used as a vehicle for holding communal investments in DH or community energy efficiency projects (heat stores, distributed generation plant).[8] Similar mechanisms might be applicable in other countries too.

Analogous contractual commitments to repay – which provide the investor with surety that they will be repaid – can be applied at the household level. The UK is introducing legislation to attach charges for energy efficiency improvements to be added to the energy bill and attached to the property, not the individual. In the US and Toronto, the PACE charge is attached to the property tax for solar panels and external insulation.

But these are just vehicles for charging beneficiaries for community energy infrastructure. Who will provide the hundreds of billions of capital itself? Curiously governments – large net borrowers – also act as a conduit for savings. Local governments, universities and hospitals have established substantial pension schemes for their staff's retirement savings. These funds hold investments of many thousand dollars/pounds/euros per employee. Governments could also guarantee loans against risks that the investor, contractor or consumer could not be held responsible. Or it could establish investment banks capitalized with public funds to lend to community infrastructure projects that are too small to raise bonds themselves. Government or local authorities could also aggregate small community energy projects to a size large enough that they can raise finance from capital markets.

Central governments should introduce a high and escalating carbon tax to significantly increase the cost of energy use

For most households, domestic energy makes up less than 5 per cent of the weekly budget, and is an even smaller proportion of most businesses' costs. Most people are not aware of how much the spend on energy; few businesses need to think about their energy efficiency. For a minority of businesses and around a quarter of households in the UK this is not the case, but these are the exception rather than the rule.

As buildings and production processes become more energy efficient, the cost of energy use falls and people can become more careless with conservation – the so-called rebound effect. This means the tax needs to be increased at the same rate as efficiency gains to stop energy use 'rebounding' back.

We argue for a high and growing tax on carbon to promote energy efficiency and transition to low-carbon energy sources. The levels presently faced by domestic consumers in Denmark of around €150/tCO$_2$ need to be implemented in other countries swiftly. A carbon market could potentially act as a substitute for a carbon tax. The issue isn't so much whether it is a tax or an emissions permit but whether it creates certainty to investors. A carbon market with a high and increasing floor price for carbon might achieve this. The European ETS has been too volatile and the carbon price too low to achieve what is needed.

A carbon tax of \$150/tCO$_2$ would raise over \$3000 per household in Western Europe and over \$6000 in North America. It is important that some of this revenue is returned to energy-intensive traded manufacturing businesses (such as ceramics, aluminium and steel), and that grants are provided to subsidize fuel-poor households' insulation and increase their income.

A tax of $150/tCO$_2$ seems a lot, but the present rate of fuel duties on transport fuels in Europe is equivalent to $350/tCO$_2$. Transportation taxes are $50/tCO$_2$ even in US. People are used to such 'energy' taxes being used to raise revenue and increasing the prices of goods in the economy. A carbon tax of $150/tCO$_2$ would add around 3c/kWh to the cost of gas and around 6c/kWh to the price of electricity. In both cases this is around half the retail price of the energy.

Incentives have an effect on consumers too; they are a form of communication. Taxes on smoking and alcohol convey society's disapproval of these goods. A pre-announced policy of rising energy taxes also changes expectations. Analysts and planners assessing different projects will change assumptions in their spreadsheets to make energy-saving projects more viable.

We think this change in policy needs to be done swiftly – an energy-efficient economy needs a strong and consistent price signal. Our economies also need more tax revenue. Such a tax would raise around 10 per cent of the government spending in the UK and would address a large part of the structural deficit faced by European countries and the US. Such a tax becomes increasingly necessary as we begin to use zero-tax electricity instead of highly taxed petrol and diesel to power our transport.

Renewable energy and CHP need incentives such as FiTs

The cost pressures facing renewable energy and CHP are different to those facing fossil fuel generation. They are exposed to price risks such as the cost of biomass, or the prices of photovoltaic modules and wind turbines. Perversely, they are also affected by carbon and gas prices because the price at which they sell electricity is set by gas- and coal-fired power plants. Renewable energy and CHP have high capital costs and lower operating costs, so are vulnerable if their assets are underutilized or if interest rates rise.

This is very different to a gas-fired power plant, where capital costs are comparatively low and the operating costs of purchased gas high. As a result, investors see less risk in financing fossil fuel power and are much more comfortable with these investments. Fossil fuel power is therefore preferred because it is a less risky investment than renewables or CHP.

The energy markets are not designed for handling small volumes of power – regardless of their cost – and punish intermittent generation. They do not always incentivize potentially responsive generation by CHP. We think FiTs should be the norm for renewable and CHP electricity to remove many of these price risks. These are operated in Germany and many other countries. This will provide consumers more stable energy prices and also greater stability of income to non-fossil fuel power plants.

Investment in DH networks needs strong regulations like the Danish Heat Law

We also recommend that local governments be allowed to mandate DH areas as they do in Denmark. In these zones, gas central heating should be phased out and replaced with mandatory connection to locally owned heat networks within an agreed period of years.

Tariff structures should be tiered to reward low energy-using households, reducing excessive and discretionary consumption by the households

There is a substantial variation between how much different households spend on electricity and gas. Often this arises from simple choices people make about what temperature to heat their homes to or whether or not to switch appliances off. Energy is often simply wasted. One idea, tried in California, is to have tiered tariffs for electricity, where the first units of electricity used are cheap – since they represent a basic entitlement. The price of electricity used beyond this entitlement increases sharply. Tiered tariffs should also be used for gas bills.

Such innovative tariff structures would work alongside behavioural nudges to provide people with more up-to-date and salient information on their energy use, and that of their neighbours.

Governments should regularly strengthen (and then enforce) energy performance standards for appliances, and energy labelling needs to be regularly updated

Much of our energy consumption arises because of the way our appliances and building components are designed and work. It is therefore intrinsically wound up in decisions about which components are selected, which optional features to include and quality control in assembly. Innovation in the power management of these goods is almost entirely driven by government standards. Few consumers have the time or inclination to understand the power consumption characteristics of their products.

Appliances are often designed with international, rather than national or local markets in mind. In Europe, appliances use the A to G energy labels to indicate performance; in the US there are also minimum federal standards and the Energy Star label for superior performance. However, in both cases, industry lobbies dominate the setting and review of regulatory standards. Progress has been swifter in the US over the past two years, and the same needs to happen in Europe.

We recommend a much more aggressive approach to the setting, testing and advertising of new standards for appliances. It is unrealistic for communities to directly participate in this, but it's important that the user voice is heard by testing how the appliance performs in the field and incorporating the feedback from users to modify design and rewrite instructions; real human beings might not work out how to use fiddly switches, might misunderstand instructions, might fail to understand or have any use for unnecessary energy-gobbling features that could be designed out of the product.

Within a few years appliances will need to communicate with the smart grid, and perhaps remotely with their owners over the internet. The protocols that govern these information flows need to be developed in conjunction with users and national governments.

Improving the energy efficiency of existing and new buildings relies on correctly selecting, installing and using building materials and products, which again need to conform to good energy labelling and customer operability practices. This list includes products as banal as draughtproof fittings on letter flaps and insulating foams that can be accurately applied, to complex new products such as heat meters and heat exchangers, solar hot water panels and zonal heating controls, just to name a few. There are huge opportunities to save energy from the correct use of innovative new products like these, but this will only happen if there is a trusted process for evaluating their performance in the field. To do this well is expensive. The results should be shared, not just within countries, but also between countries. We think this is a role for government or consortia of the energy service companies described above, perhaps funded by levies on the manufacturers.

Community ESCOs should reduce energy use by targeting customer behaviour by providing salient information, nudges and social pressures

People and businesses, including those sympathetic to sustainable energy, find it difficult linking their actions to their beliefs. There is a substantial gulf in what they say they care about and what they do. Humans are social creatures and imperfect consumers. Behavioural change programmes could harness these two quirks for public purpose.

There is much a community ESCO can do to harness people's appetite to reduce their own energy use. There are many programmes that show that feedback providing social comparison information on bills or energy monitors can make a difference, especially to families that use abnormally high amounts of energy.

People respond to social pressures. If they make a commitment, they will try and stick to it. A simple pledge to undertake a small number of changes to their behaviour can, without sanction or pressure, motivate people to bring forward long-promised changes. People are competitive and don't like disappointing their friends and neighbours. The *Energy Smackdown* contest in Massachusetts shows how communities can harness the competitive impulse. Using a game show format, different family or neighbourhood-based teams compete over several episodes to learn and apply techniques to reduce their home's energy and imported food use.

People's behaviour can be influenced by watching and learning from others, and working together for a common goal. These 'communities' may include neighbours, family, colleagues or local leaders. Community-based dissemination can be useful for teaching or exemplifying small changes to their everyday lives: how to set the heating controls correctly, how to improve a home's airtightness by installing draughtproofing measures, and low-energy lighting.

People like working in groups. The Transition Towns network brings people with a shared concern about peak oil and climate change together and marshals their enthusiasm into a practical and radical agenda for community action. The movement has adopted a franchise model to spread. Groups are largely independent of one another – a decentralized, distributed system of interaction – but knowledge is freely disseminated: teaching materials, branding, the programme of actions. These independent groups are now connecting with formal local structures, working on contracts with local government to provide training for low-carbon retrofit or teach local food production.

Simple *nudges* can be used to make low-energy behaviour the default option. Local builders are responsible for many of the decisions about which fittings and materials to use. While going to work on one job they can educate their clients about low-energy loans or grants for installing energy-efficient changes: why not install underfloor insulation while the house is being rewired? The floorboards have been raised anyway and government is offering a low-interest loan to cover the extra cost.

Local government and community ESCOs must ensure the local workforce is trained and accredited

New standards such as Passivhaus in Europe and Leadership in Energy and Environmental Design in North America set out the standards of performance for new construction. Very few homes have been built to these demanding standards. Industry does not have the appropriate skills and at present it is not required to learn. Even the building regulations that governments set, which are much less efficient than Passivhaus, are often left unenforced.

Many of the techniques and detailed construction practices also need to be applied to retrofitting existing homes. These skills have to be diffused through the workforce. By acting as major local customers, local government and community ESCOs can train workers with the correct professional skills and a working knowledge of building science concepts, such as making buildings more airtight to stop uncontrolled losses of heat and controlled ventilation with heat recovery to ensure buildings don't become damp.

Businesses will train workers if they are confident the new standards are stable, and that the inspection regime will detect and punish poor performance. Well-trained fitters are effective ambassadors and trusted salesmen of new technologies or measures. Good practices in energy efficiency are often hard to detect and rely on effective and well-enforced regulation. It is important for building up consumer confidence that local governments' building inspectors are independent of contractors and sufficiently resourced to ensure compliance.

All This Has to Be Done Now

We are not arguing that repowered communities should entirely replace centralized solutions. Countries have huge existing investments in large power stations, transmission lines, international power and gas flows. We will continue to need substantial investment in large power plants: coal and gas with CCS, offshore wind and tidal to meet the increased need for power as road transport electrifies and heat pumps are used outside DH areas.

But we are arguing that community solutions be encouraged and that centralized solutions should be made subordinate to community energy. Community energy and energy efficiency will create jobs, save money and make the energy system as a whole more secure. Too much policy effort is being put into servicing the needs of large power companies. Policy-making seems to be holding up its hands, surrendering to the idea that there are no alternatives.

We hope this book shows readers that there is an alternative. It already exists in pockets through North America and Europe. No country or community has got everything right. And even if it had, that situation could not be cloned and planted into another soil. The examples of good practice covered in this book are still the exception rather than the rule. But much greater effort must be put into understanding what has worked and then applying it to other places. Ownership of our energy challenges has to transfer from companies to communities. It is through the efforts and understandings of everyday people that we will reduce demand and accept more sustainable energy use.

Is it possible to do all this? We think it is. The largest urban development project in Europe at the time of writing is the creation of Olympic Park in London. In a few short years a town of sports arenas, accommodation for athletes and shops is being built. The French-owned energy company Cofely is installing two 20MW tri-gen power stations that provide power, heating and cooling for the site. Sixteen kilometres of DH and cooling pipes convey low-carbon heat and cool around the site. It has been possible to finance the project because it carries a 40-year build and operate concession. Within the 300 hectares of land that makes up the Olympic Park site, all developments have to connect to the DH system. Over the life of the DH network, the fuel may well change from gas to biomass and waste, or even surplus heat from a carbon-abated fossil fuel plant, as the economic incentives for carbon and fossil fuel prices alter. The Olympic

Delivery Authority has trained many hundreds of unemployed people from the four local government boroughs to provide the workforce on the site.

At the opening ceremony for the energy centre, the UK's energy minister Greg Barker quipped: '… unlocking the potential of decentralized energy will take an Olympic effort'.[9] He is right. We hope this book has shown the range of events policy-makers have to get right and – equally importantly – that like the Olympics itself, competition occurs best within a framework of friendly co-operation.

Notes

1   Megginson, L. C. (1963) 'Lessons from Europe for American Business', *Southwestern Social Science Quarterly*, vol 44, no 1, p4.
2   Fleming, D. (2007) *Energy and the Common Purpose*, 3rd edition, The Lean Economy Connection.
3   *The Economist*, 11 March 2004, www.economist.com/node/2476988.
4   Margulis, L. (1970) *Origin of Eukaryotic Cells*, Yale University Press.
5   Song lyrics; see www.myspace.com/projectearthrock1.
6   Carbon Finance (2010) 'Utility-only cap and trade no silver bullet', www.carbon-financeonline.com/index.cfm?section=lead&action=view&id=13042.
7   Plugincars (2010) 'Exclusive Mercedes interview: Electric cars are not the silver bullet', www.plugincars.com/exclusive-mercedes-interview-evs-are-not-silver-bullet-87841.html.
8   Vaze, P. and Mayo, E. (2009) *A New Energy Infrastructure*, Consumer Focus, www.consumerfocus.org.uk/assets/1/files/2009/06/A-New-Energy-Infrastructure2.pdf.
9   Unpublished speech given by Greg Barker at Olympic Park Energy Centre on 25 October 2010.

Glossary

20/20/20 package. The EU agreement that greenhouse gas emissions must be reduced by 20 per cent, 20 per cent of all energy must come from renewables and there must be a 20 per cent improvement in energy efficiency, all by 2020.

Adaptation. Steps to prepare to deal with the now-unavoidable impacts of climate change, such as sea-level rise.

Anaerobic digestion. Technology that converts organic waste (food, farm waste, sewage) into compost and renewable gas (also known as biogas).

Bioenergy. Energy derived from plants or other organic material such as food, farm or human waste. Plants used to generate electricity are referred to as biomass, plants used for transport fuel as biofuels and gas from organic waste as biogas.

Cap-and-trade. Regulatory system that sets a total limit for the amount of pollution that can be emitted by all the sectors covered but then allows the companies within those sectors to trade permits. Cap-and-trade was used by US states to reduce acid rain pollutants. The EU-ETS was the first international cap-and-trade scheme for greenhouse gases.

Carbon capture and storage (know as sequestration in the US). Technology that captures CO_2 either before a fossil fuel is combusted, during combustion or after combustion. The CO_2 is then transported – either by pipeline or by truck/ship and stored in old oil or gas fields (which leads to enhanced oil or gas recovery) or in saltwater aquifers.

Combined heat and power (also known as cogeneration). Technology that means that when anything is burnt to generate electricity, the heat is also used, for either industrial or domestic heating.

Decentralized energy. An approach in which electricity is generated in many small-scale power plants rather than a few large ones. Decentralized energy makes CHP much more practicable, because heat is more difficult to transport than electricity.

Demand-side management. Regulatory system requiring energy companies to spend money on energy efficiency programmes.

District heating. The transport of heat from a centralized heat source (often a CHP plant) to homes, offices and factories. DH is thus an alternative to each property having its own boiler.

Emissions performance standard. Regulation limiting the amount of greenhouse gas that a power station can emit per unit of electricity generated.

Energy intensity. A way to measure the energy efficiency of an economy by measuring the amount of energy used per unit of GDP.

Energy security. The amount of energy that a country can get from within its own territory or territorial waters, or (more loosely) from friendly countries.

Energy services company. A company that sells services such as keeping a property warm or cool, and is paid if the promised outcome is achieved, rather than being paid on the basis of amount of fuel used.

Environmental tax reform. Shifting taxes from 'goods' – income, employment – and onto 'bads' – pollution, waste, resource depletion.

Feed-in tariff. A system offering those who produce desirable forms of energy (renewables or CHP) a guaranteed level of income for the energy, at a higher level than the tariff for fossil fuel energy.

Fuel poverty (also known as energy poverty). The inability to afford enough energy to keep warm or cool. The UK has an official definition of fuel poverty: when more than 10 per cent of disposable income has to be spent on fuel.

Heat exchanger. Device used to transfer heat or coldness between fluids. For instance, warm stale air has its heat recovered through a heat exchanger before it is expelled.

Light emitting diodes. Very energy-efficient light bulbs.

Liquefied petroleum gas. Gas turned into a liquid. This is either from natural gas or from oil.

Low-carbon energy. Energy that does not result in significant greenhouse gas emissions. Gas emits only 42 per cent of the carbon emissions of coal without CCS, nuclear 12.5 per cent, solar photovoltaic 11.5 per cent, coal with CCS 9.5 per cent and wind 2.5 per cent. See http://climateanswers.info/2009/12/how-low-carbon-are-different-generating-technologies.

Mitigation. Measures taken to reduce greenhouse gas emissions.

Net metering. A requirement on the electricity network operator to pay for renewable electricity exported to the grid. This is similar to a FiT, though the rate is the same as for 'brown' electricity and so not nearly as remunerative in promoting renewables.

Regulated asset base. Regulatory system in which the regulator agrees with companies' investment in new assets and allows the companies to seek finance. The finance can then be borrowed at very cheap rates as the income is guaranteed from the customers.

Renewables Obligation/Renewables Portfolio Standard. An obligation on electricity supply companies that a certain proportion of their electricity comes from renewables.

Retrofitting. Improving the energy efficiency of an existing building.

Rising block tariffs. Under these, households' first units of energy consumption are cheap but the next block of consumption is more expensive.

Run-of-river hydro. A means of using a river to generate electricity without building a dam, by simply submerging a turbine into the running water.

Shale gas. Gas from unconventional sources such as using techniques like squirting liquid into rocks at very high pressure to crack the rock. This might increase the supply of gas substantially, but is expensive, may harm nearby groundwater and may have high life-cycle greenhouse gas emissions.

Social discounts. Mandatory subsidies that energy companies have to offer to vulnerable households.

Solar energy. Energy from the sun can be used to generate electricity via photovoltaic panels. Sunlight can also be used to boil water, if concentrated via mirrors, with the steam then turning a turbine and generating electricity. This is called concentrated solar power. The sun can also be used to provide space and water heating – solar thermal.

Thermal imager. Camera that takes pictures of buildings and shows where heat is escaping – usually in bright colours.

Time-of-use tariffs. These vary the price of electricity according to the time of the day to encourage people to use electricity when it is most in surplus (such as during the middle of the night).

Transition Towns. Network of groups aiming to help communities make the transition from a high-energy consumption town to a low-energy town. Transition towns originated in 2005 in Ireland. There are now over 300 official Transition initiatives, about half of which are in the UK. The US has developed its own network based on the European model, involving 75 local initiatives.

Weatherization. US term for energy efficiency work.

Index